•

The quest for the farthest objects in the Universe remains one of the most challenging to modern astronomy. Peering deeper and deeper into space reveals the most distant and powerful objects known and so probes back to the embryonic epochs of the Universe not long after its birth in the Big Bang.

Four world experts – chosen for their ability to communicate research astronomy to popular audiences – each contribute a chapter to this lucid survey. In clear terms they bring to the general audience the excitement and challenge of studying the Universe on the largest scales. They address the fundamental issues of scale in the Universe; the ghostly etchings seen on the cosmic background radiation; quasars and their evolution; and galaxy birth.

This survey offers an exceptional chance for the general audience to share in the excitement of today's forefront research of the early Universe in an accessible and stimulating way.

•

·

THE
FARTHEST
THINGS
IN THE
UNIVERSE

·

THE FARTHEST THINGS IN THE UNIVERSE

JAY M. PASACHOFF

WILLIAMS COLLEGE

HYRON SPINRAD

UNIVERSITY OF CALIFORNIA AT BERKELEY

PATRICK S. OSMER

THE OHIO STATE UNIVERSITY

EDWARD S. CHENG

NASA – GODDARD SPACE FLIGHT CENTER

CAMBRIDGE
UNIVERSITY PRESS

Published by the Press Syndicate of the University of Cambridge
The Pitt Building, Trumpington Street, Cambridge CB2 1RP
40 West 20th Street, New York, NY 10011–4211, USA
10 Stamford Road, Oakleigh, Melbourne 3166, Australia

First published 1994

Printed in Great Britain at the University Press, Cambridge

A catalogue record for this book is available from the British Library

Library of Congress cataloguing in publication data

The Farthest things in the universe / Jay M. Pasachoff . . . [*et al.*].
p. cm.
Includes index.
ISBN 0-521-45170-1. – ISBN 0-521-46931-7 (pbk.)
1. Astronomy. 2. Cosmology. I. Pasachoff, Jay M.
QB43.2.F37 1994
520–dc20 94–4679 CIP

ISBN 0 521 45170 1 hardback
ISBN 0 521 46931 7 paperback

CONTENTS

PREFACE

This book originated as a symposium at the American Association for the Advancement of Science annual meeting in San Francisco in 1989. The topic, *The Farthest Things in the Universe,* suggested itself to me as the most interesting and significant topic that people could hear about. An earlier AAAS Symposium had led to a book, *The Redshift Controversy,* that was still in use, and we hope that this volume will prove itself of similarly lasting interest.

Two of the original speakers, Hyron Spinrad of the University of California at Berkeley, and Patrick Osmer, then of the National Optical Astronomy Observatories, revised their pieces to bring them up-to-date for inclusion in this book. Further, Ed Cheng of the COBE Science Team and NASA's Goddard Space Flight Center agreed to write a new piece for inclusion in the book. We appreciate his taking time during the period of his duties as Chief Scientist for the Hubble Space Telescope's repair mission to complete his piece. During the interval from the time of the symposium to the present, the Cosmic Background Explorer spacecraft was launched and has had its tremendous successes in showing that the Universe has a black-body spectrum and in finding ripples in space that may be the seeds from which galaxy-formation began. Thus this book appears at an optimum time.

During the period of preparation of the book, I was resident for a year at the Institute for Advanced Study at Princeton, and I thank John Bahcall for his hospitality there. In Williamstown, I have been assisted by Susan Kaufman. The book was completed during a sabbatical leave at the Harvard-Smithsonian Center for Astrophysics, and I thank Harvey Tananbaum for his hospitality there.

Ed Cheng thanks Mike Hauser, Matt Kowitt, John Mather, Steve Shore, and Rai Weiss for their helpful and critical comments on the various drafts of the manuscript. Any remaining inaccuracies are, of course, a consequence of his persistent blockheadedness. He especially thanks the COBE team of scientists, engineers, and computing experts for showing what teamwork, in the best NASA tradition, can accomplish.

Hy Spinrad thanks especially the staffs of the Kitt Peak National Observatory and of the Lick Observatory for much cooperation over the years. He again acknowledges the help of his colleagues, students and recent ex-students for keeping this research going over a long and productive decade. In particular he thanks Wil van Breugel, Pat McCarthy, and Mark Dickinson. Much of his research on galaxies has been supported by the U.S. National Science Foundation – and he thanks them, also.

Patrick Osmer thanks the National Optical Astronomy Observatories (NOAO) and their staff for support of his research and of the writing of his contribution. NOAO is operated by the Association of Universities for Research in Astronomy, Inc., under contract with the National Science Foundation. He is particularly grateful to his collaborators Paul Hewett and Stephen Warren for discussions on the search for high-redshift quasars.

We are grateful to Nancy Kutner for the index. We all thank Simon Mitton and Adam Black of the Cambridge University Press for their interest in the book and for bringing it to completion.

Jay M. Pasachoff
Williamstown, Massachusetts

CHAPTER ONE

•

OBSERVING THE FARTHEST THINGS
IN THE UNIVERSE

•

JAY M. PASACHOFF

When we look out into space at night, we see the Moon, the planets, and the stars. The Moon is so close, only about 380 000 kilometers (240 000 miles) that we can send humans out to walk on it, as we did in the brief glorious period from 1969 to 1972. Even the planets are close enough that we can send space-craft out to them, notably the Voyager spacecraft, one of which has passed Neptune. Whereas light and radio signals from spacecraft take only about a second to reach us from the Moon, the radio signals from Voyager 2 at Neptune took several hours to travel to waiting radio telescopes on Earth. We say that the distance to the Moon is 1 light-second and the distance to Neptune is several light-hours.

Aside from our Sun, the nearest star at 8 light-minutes away, the distances to the stars are measured in light-years. The nearest star system is Alpha Centauri, visible only in the southern sky, and the single nearest star is known as Proxima Centauri, about 4.2 light-years away. We know so little about the stars that new evidence in 1993 indicates that Proxima Centauri might not be a member of a triple-star system along with the other parts of alpha Centauri, as has long been thought. The speeds at which those stars are moving through space may be sufficiently different that Proxima is only tem-porarily near Alpha's components.

MEASURING DISTANCES IN THE UNIVERSE

Only for the nearest stars, those within about 100 light-years of our Earth and Sun, can we find their distances by a fairly direct method. This method depends on the concept of parallax, in which objects seem to shift with

1

respect to a background when we look at them from different point of view. The easiest way to demonstrate parallax is to hold out your arm and to look at your thumb with first only your left eye and then only your right. You will notice that your thumb seems to jump with respect to the distant wall as you alternately blink your eyes. If you hold your thumb closer, the parallax from eye to eye is greater. This parallax effect also shows up in automobile speedometers, as the needle looks to be in slightly different places when seen from the point of view of the driver and of the passenger.

Parallaxes for stars have only been measured since 1838, since they are so small. We measure them from different points of view, just as we do with left eye/right eye, but for stars the different points of view are the Earth's orbit at intervals of several months. The points of view are thus, at best, twice the distance to the Sun, which is twice 150 million kilometers (93 million miles). Except for Alpha Centauri and Proxima, the parallaxes we measure are all less than 1 second of arc. One second of arc is a tiny angle, 1/60 of a minute of arc and 1/3600 of a degree of arc. Since the Moon is only half a degree, or about 30 minutes of arc, across, one second of arc is about 1/20 of one per cent of the apparent diameter of the Moon. Our unaided eyes can distinguish angles of about 1 minute of arc, so it take a telescope to measure the parallax to even the nearest stars.

A European spacecraft has been in orbit for the last few years measuring parallaxes to unprecedented precision. This Hipparcos spacecraft (for High Precision Parallax Collecting Spacecraft, a pun on the name of the Greek astronomer Hipparchus who measured star brightnesses two thousand years ago) didn't go into the proper orbit, but scientists and engineers have nonetheless been able to get enough good data from it to calculate parallaxes for over 100 000 stars. Since the smaller the parallax shift, the larger the distance, these newly precise values give us more-or-less accurate values for the distances to these stars. But even so, these values are for stars measured only in 10s of light-years.

Our galaxy, the Milky Way Galaxy, is an assortment of stars, gas, and dust containing about a trillion (1 000 000 000 000) times the mass of the Sun; we say that it contains 'a trillion solar masses.' (To keep track of large numbers without having to count zeroes all the time, we commonly put the number of zeroes as a superscript to the number 10: 1 trillion is 10^{12}. The system works even with small values for the exponent, since we define 10^0 as 1. Thus 10^1 is 10, 10^2 is 100, and so on.) Our Sun is located in a spiral arm of our galaxy, about 30 000 light-years from its center. Thus stars far from us in our galaxy are much too far away for us to measure their distances using parallaxes.

Instead, we often determine what types of stars they are by looking at their spectra or at the distribution of the amount of light they give off in different colors. Since we know how bright standard stars are with those properties, and we can measure how bright the star we are examining is, we figure out how far away the star has to be to appear as bright as it does. The concept is like figuring out how far away a lamppost is on a dark night by how bright it looks. Brightness goes down with the square of the distance. Thus a star that is twice as far away as another star of the same intrinsic brightness appears one-half squared, or one-fourth, as bright.

The nearest galaxies are a pair of small satellite galaxies to our Milky Way. Since they were brought to the attention of European astronomers by the crew of Magellan's ship, who noticed them when they got sufficiently far south, they are known as the Magellanic Clouds. They are only about 400 000 light-years away. Astronomy got a big boost and lots of excitement in 1987 when a star exploded in the Large Magellanic Cloud. This supernova became bright enough to be seen on Earth with the naked eye; since it is in another galaxy, we can calculate that it had to be extremely bright, about as bright as all the rest of the stars in that galaxy put together. Supernova 1987A, as it is known, is still being monitored, though it has faded enough so that it can be observed only with telescopes.

The next galaxy like our own, moving outward from us, is the Great Galaxy in Andromeda (Fig. 1.1). It is known as M31, from its position as the 31st object in a catalogue of non-stellar objects compiled by Charles Messier in France over two hundred years ago. Messier was compiling a list of objects to avoid confusing them with the comets he was searching for, but his list turns out to contain about 100 of the most interesting objects of the sky. Whenever an object has a Messier number, we usually use it, and it broadcasts that 'here is an object pretty bright and easy to observe.' The Andromeda Galaxy is a spiral galaxy some 2.2 million light-years away and is perhaps the farthest object we can see with the unaided eye. Another spiral galaxy, M33 in the constellation Triangulum, is slightly farther away and can also be seen by some people with unaided eye.

How do we tell distances to these galaxies, which are all much too far away for us to measure parallaxes? Indeed the problem plagued astronomy for a long time, especially since in the early years of this century a distinguished astronomer reported that he had measured some motion in M31, akin to a small parallax as an indicator that something is very close by. The matter was resolved only in the 1920s by Edwin Hubble, a Rhodes scholar and lawyer turned astronomer, who used the then largest telescope in the

Fig. 1.1. The Great Galaxy in Adromeda, M31, a spiral galaxy containing a trillion times the mass of the Sun. It is seen obliquely. (Palomar Observatory photograph)

world, the 2.5-m (100-inch) reflector on Mt. Wilson in California, to measure the brightnesses of several variable stars in M31, the Andromeda Galaxy. (Of course, there are really thousands of galaxies we can detect with a large telescope in the constellation Andromeda, but 'the Andromeda Galaxy' is generally understood to mean M31.)

Earlier, Henrietta Leavitt at the Harvard College Observatory had found the key in observing a type of variable star known as Cepheid. They got their name from their prototype, the star delta Cephei (that is, the fourth brightest star in the constellation Cepheus), which gets brighter and fainter by a small factor with a period of a few days. Leavitt discovered that the longer the period, the brighter the star. So by simply following how long a Cepheid

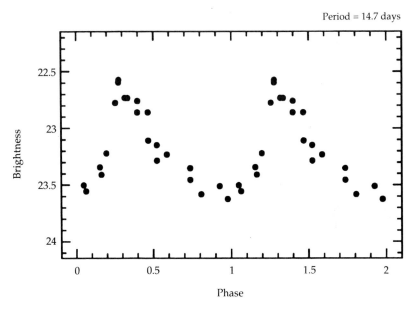

Fig. 1.2. A light curve for one of the 31 Cepheids studied in the galaxy M81 as part of one of the Key Projects of the Hubble Space Telescope. The horizontal axis shows the phase of the 14.7-day cycle (a cycle is repeated here) while the vertical axis shows the magnitude (brightness) in the spectral band known as V, a yellow band that corresponds to the visible part of the spectrum best seen with the eye; 23rd magnitude is very faint, fainter than ground-based telescopes can observe at such high accuracy. (Wendy L. Freedman, Carnegie Observatories; Shaun M. Hughes, Barry F. Madore, and Jeremy R. Mould, Caltech; Myung Gyoon Lee, Carnegie Observatories; Peter B. Stetson, Dominion Astrophysical Observatory; Robert C. Kennicutt and Anne Turner, Steward Observatory; Laura Ferrarese and Holland Ford, Space Telescope Science Institute; and John A. Graham, Carnegie Institution of Washington)

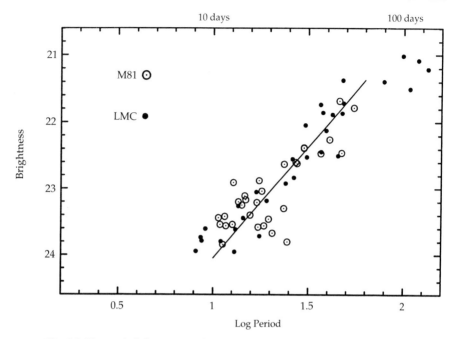

Fig. 1.3. The period–luminosity diagram, graphing the period in days vs. the (V) magnitude of the stars (on a logarithmic scale of period, with 1=10 days, 1.5=32 days, and 2=100 days). Open circles show the Hubble Space Telescope measurements of the galaxy M81, while filled circles show Cepheids in the Large Magellanic Cloud (LMC) shifted by an amount that corresponds to the difference in distances of the two galaxies from us (9.09 magnitudes). (Freedman, *et al.*, as in previous figure)

variable star takes to go through a cycle of brightness, we can know how bright it is intrinsically. Then we can use the standard method of comparing how bright it really is with how bright it looks to find out how far away it is. Leavitt worked out her method for a nearby galaxy, the Large Magellanic Cloud, and Hubble's extension of the method to the Andromeda Galaxy gave us the distance to unprecedented accuracy. It was this observation that showed, indeed, that M31 was a distant galaxy like our own and not merely a spiral cloud of gas in our own galaxy.

The method is still our best way of finding distances to objects reasonably far away. One of the major goals of the Hubble Space Telescope is to use its unique capabilities to detect Cepheid variables in galaxies as far away as possible, as part of the accurate determination of the cosmic distance scale. The first step in that important task was reported in 1993 with the determination of the distance to the spiral galaxy M81 from studies of 30 newly discovered

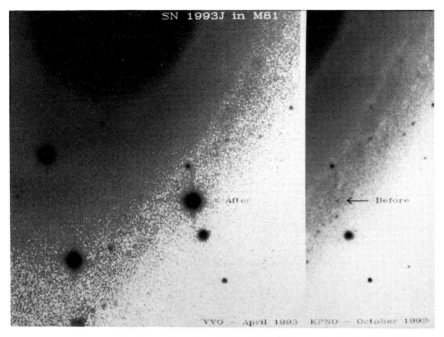

Fig. 1.4. The spiral galaxy M81 in Ursa Major, both before and after its March 28, 1993, eruption of Supernova 1993J. (John Salzer, Wesleyan University)

Cepheids in it with periods from 10 days to 55 days (Figs. 1.2, 1.3). M81 turns out to be 11 million light-years away, a distance determined through this method by the Hubble Space Telescope to plus or minus 10 per cent, a much lower uncertainty than previous measurements had. The work was carried out by Wendy Freedman of the Observatories of the Carnegie Institution of Washington (which formerly ran the Mt. Wilson Observatory and worked jointly with the Palomar Observatory, but which now concentrates on observations with its own telescope in Chile and on other astronomical research) and colleagues from many institutions.

The determination of the distance to M81 had immediate applicability, since a supernova was discovered in it. This supernova (Fig. 1.4) erupted on March 28, 1993, and grew to be the brightest supernova visible from Earth's northern hemisphere observatories since 1937, over 50 years. Since the intrinsic brightness of an object can be determined by measuring its apparent brightness and then scaling with the inverse-square law for its distance, the analysis gave us a more accurate brightness for this type of supernova than we previously had. Since supernovae are much brighter than Cepheid variables, knowing more about supernovae enables us to find distances to

7

more distant galaxies in which supernovae are discovered. If Supernova 1993J in M81 had turned out to be the kind of supernova that comes from the incineration of a white dwarf star when it reached the maximum mass that such stars can have, we would then have found the 'standard candle' that astronomers want to measure distances throughout the Universe. But Supernova 1993J seems to come, instead, from the collapse of a massive star. These stars collapse and reach a wide variety of brightnesses, so they can help determine distances but not as accurately.

THE EXPANSION OF THE UNIVERSE

Hubble, also in the 1920s, analyzed the spectra of dozens of galaxies. The farther away the galaxy, the longer the exposure he had to take on film, even with the world's largest telescopes. Some of his exposures were 16 hours long or longer; he would have to shut the telescope down at the end of the night and resume the next night. We shall see later on, that such exposures can now be obtained in minutes, using today's electronic detectors.

In any case, when Hubble examined the galaxy spectra, he saw that there was a tendency for the farther galaxies to have their spectra shifted to longer wavelengths by a greater factor than nearer galaxies (Fig. 1.5). Thus the particular traces of calcium in the spectrum of the galaxy, a feature that is particularly strong and easy to notice, appeared not quite as far into the

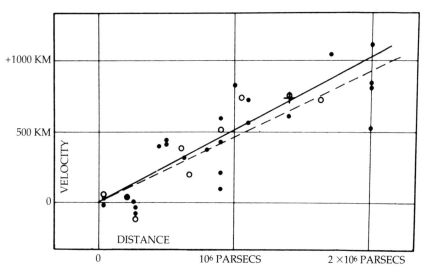

Fig. 1.5. The 1929 Hubble-law diagram compiled by Hubble. (National Academy of Sciences)

ultraviolet as they did on Earth. Since they were slightly moved to a direction toward the red from the ultraviolet, they are 'redshifted,' even though they don't actually turn red. Hubble made the leap of genius and suggested that there was a direct link between the amount of redshift and the distance of the galaxy. In future years, he worked with Milton Humason at Mt. Wilson to push the data farther out into the Universe, and his extrapolation proved correct. The relation between distance and redshift was a straight-line proportionality as far as he could determine. Indeed, the relation still seems true, and we even now argue and debate about a possible tiny curvature at the top end of the straight line.

This Hubble law of expansion, the law linking redshift and distance, is the main way we determine distances to the most distant galaxies. But it is based on observations for nearer objects, in which we can determine the distance in some other way. We try now to measure as many of these distances to as high an accuracy as possible.

The conundrum that astronomy has had for decades is that different groups of people measure different values for the Hubble constant, the constant of proportionality linking velocity and distance: $v=H_0 d$, where v is velocity, d is distance, and H_0 is the Hubble constant (or 'Hubble's constant'), with the subscript zero indicating that the Hubble constant is measured at some beginning time. The traditional value measured for some decades by Allan Sandage, Hubble's heir at the Mt. Wilson and Palomar Observatories, and his colleague Gustav Tammann, is some 50 kilometers per second per megaparsec. Some other scientists found 100. But other groups are now often finding a value closer to 150. The debate is still going on hot and heavy. We look for the Hubble Space Telescope and some of the new generation of ground-based telescopes to resolve the controversy.

What do the strange units of Hubble's constant mean? A parsec is a way that astronomers use to measure distance, since it is easily computed by taking one over the measured parallax angle. If you are at a distant star and look back at the radius of the Earth's orbit, it takes up ('subtends') a certain angle in the sky. If you go far enough back, it takes up about a degree; if you go about 60 times farther back, it takes up a minute of arc; and if you go another 60 times farther back, it takes up a second of arc. At this distance, you are one parsec from Earth. The distance works out to be about 3.3 light-years. Now we can analyze Hubble's constant. For each million parsecs (megaparsec, or Mpc) you go away, the speed at which a galaxy is receding increases by 50 kilometers per second or 100 kilometers per second, depending on which group is correct. Thus at a distance of one million parsecs (taking the first number), a galaxy is receding at 50 kilometers a second; at two million parsecs, a galaxy is receding at 100 kilometers a second; at three million parsecs, a galaxy is receding at 150 kilometers a second, and so on. Actually, for the nearer galaxies like Andromeda, which is only 2.2 million light-years and thus less than one parsec away, the galaxy has its own motion to and fro, called 'peculiar motion,' which adds to or subtracts from the velocity from Hubble's law. So one has to go to more distant galaxies to apply Hubble's law properly.

Fig. 1.6. The expanding Universe is often likened to a rising raisin cake, with the galaxies corresponding to the raisins. An individual galaxy, like an individual raisin, is not expanding. If you were sitting on any raisin, though, you would see all the raisins receding from you. You need not be at the center of the cake to have this feeling. The Universe, indeed, may be infinite and have no center. (Jay M. Pasachoff photograph)

HUBBLE'S LAW AND THE AGE OF THE UNIVERSE

From the fact that the Universe is expanding (Fig. 1.6), we can consider what would happen if we went backward in time instead of forward. Indeed, from the rate the Universe is expanding (as measured by Hubble's constant, for example), we can figure out how fast the Universe would be contracting, and when it would all be together at infinite density. Thus we can calculate the 'age of the Universe' by simply taking the inverse, that is, one over, Hubble's constant. When all the kilometers and megaparsecs are changed into the same unit so that they can be cancelled out, one over the Hubble constant ($1/H_0$) comes out to about 20 billion years for a Hubble constant of 50 and 10 billion years for a Hubble constant of 100. So we see that the current debate over a factor of two in the value of the Hubble constant translates into an uncertainty in the age of the Universe by a factor of two.

Now, simply computing the age of the Universe in this way assumes that the Universe always has and always will expand at the current rate. It would do so only if there were no gravity pulling back on all its matter. Theorists also consider a term called 'the cosmological constant,' introduced into his equations by Albert Einstein (though it later seemed to be unnecessary, leading Einstein to say that he had made a mistake by introducing it). This cosmological constant amounts to an effect speeding up rather than slowing down the Universe's expansion. If we consider a cosmological constant of 0, which is usual, and a Universe whose expansion is slowing down because of gravity, we find an age for the Universe of two-thirds the value calculated above. So the range of ages we calculate is from about 7 to 14 billion years, depending on which Hubble constant we accept.

The difference is a serious one, because we want the Universe to be older than the things in it. And among the oldest things in the Universe are the globular clusters (Fig. 1.7), spherical groups of perhaps 100 000 stars that form a halo centered at our galaxy's core. The methods for dating globular clusters involve calculating how individual stars age and when they burn up the nuclear fuel at their cores. By comparing the properties of the stars' temperature and brightness, and graphing these properties for all the stars in a cluster in a way first done in 1911 by Ejnar Hertzsprung, we find a graph whose shape is characteristic of the distribution of ages. (Henry Norris Russell in 1913 plotted similar quantities for stars not in clusters.) The Hertzsprung–Russell diagrams for all globular clusters have about the same shape, making all these clusters the same age. They give an age that is over 10 billion years and, in some calculations, is close to 14 billion years. Since it

11

Fig. 1.7. The globular cluster M13 in the constellation Hercules, a set of 100 000 stars in our galaxy. (Palomar Observatory photograph)

takes a billion or two years for the stars in a cluster to form after the Universe formed, this exercise seems to rule out the shorter age for the Universe and is even running into problems for the longer age, especially as new research on globular clusters uncovers some range in their ages and some older examples. Still, there is no obvious flaw in the measurements of Hubble's constant that lead to the younger age. We await the resolution of the conundrum in the years to come.

FAR AWAY IN THE UNIVERSE

As we move away from the Earth to the Sun, from the Sun to the rest of the Milky Way Galaxy, and from the Milky Way Galaxy to galaxies like Andromeda and M81, we are still in the set of galaxies we call our Local Group. This Local Group is part of a larger cluster of galaxies known as the Virgo Cluster. To study galaxies in still more distant clusters was formerly very inefficient because they are so faint. As we shall see below, new methods of observing allow us in recent years to carry out observations of much greater statistical accuracy to greater distances. Maps of thousands of galaxies have been made, in particular, by Margaret Geller and John Huchra of the Harvard–Smithsonian Center for Astrophysics. Their 'slices of the Universe' (Fig. 1.8) reveal chains of structure extending over about a billion light-years, much larger structures than had been thought to exist (Fig. 1.9). As instru-

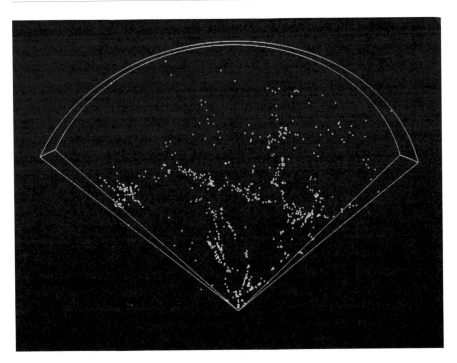

Fig. 1.8. One of the slices of the Universe for which distances were measured by John Huchra and Margaret Geller to allow this three-dimensional display of a wedge of space whose point is at the Earth. (John Huchra and Margaret Geller, Harvard–Smithsonian Center for Astrophysics)

mental capabilities improve, such mapping will extend to even larger scales.

At the distances of the distant clusters of galaxies, we find a set of objects known as quasars. When they were discovered in 1963 as sources of radio emission at the location of point-like optical objects, they were known as 'quasi-stellar radio sources,' from which 'quasar' was contracted. We date their discovery to the realization of Maarten Schmidt of the Palomar Observatory that their spectra are greatly redshifted, putting the quasars at huge distances according to Hubble's law. Three decades of studies have found thousands of quasars and have taught us much about their properties. Their distances, in any case, rely on the expansion of the Universe and Hubble's law. We now think of quasars as bright events in the cores of distant galaxies, and sometimes find them associated with clusters of galaxies. For the quasars that are not associated with clusters of galaxies, it may be that the quasars are bright enough to be detected from Earth while the other galaxies are merely too faint. The favored model for providing the energy for

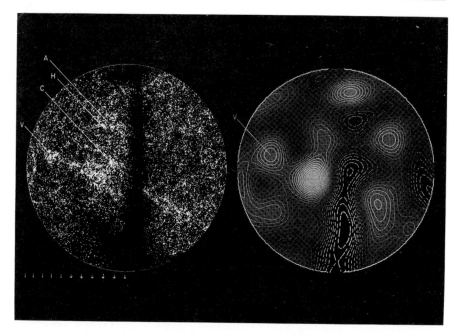

Fig. 1.9. Thousands of galaxies displayed in a half-sky map display clumps and filaments. The large clump to left of center (C) is the Centaurus Cluster of Galaxies and the clump farther to the left is the Virgo Cluster (V). That they share a common motion is a sign of the massive Great Attractor beyond that may be pulling galaxies toward it, distorting the Hubble flow. The letters A and H mark the Antlia and Hydra Clusters of Galaxies, respectively. The vertical band is the obscuration by the Milky Way. The figure on the left shows optical results, while the figure on the right shows infrared results from the Infrared Astronomical Observatory (IRAS) satellite. (*left:* David Burstein, Sandra Faber, Roger Davies, Alan Dressler, D. Lynden-Bell, Roberto Terlevich, and Gary Wegner; image by Ofer Lahaf, Institute of Astronomy, University of Cambridge; *right:* Caleb Scharf, Institute of Astronomy, University of Cambridge, courtesy of D. Lynden-Bell of the IoA and of Jacqueline Mitton for the Royal Astronomical Society)

quasars to shine so brightly is the presence of a giant black hole containing millions of times the mass of the Sun. As gas from the rest of the galaxy surounding the quasar, or, in many cases, from a colliding galaxy, falls into the black hole, it is heated to tremendous temperature and radiates strongly in all parts of the spectrum.

Only two years after the first quasars were identified, Arno Penzias and Robert Wilson of what was then the Bell Telephone Laboratories detected a faint radio hiss coming equally as a background from all directions in space.

When this observation was teamed up with theoretical work by Robert Dicke, P. James Peebles, David Wilkinson, and others of Princeton, it was realized that this background radiation was a fundamental property of the Universe. (Indeed, it had been predicted in the 1940s by George Gamow and Ralph Alpher.) In some sense, this background radiation that we are now seeing is the farthest observable thing in the Universe, since the radiation we are now receiving has been travelling through space to us for only a million years or so less than the age of the Universe. But we are bathed in this background radiation, which has been redshifted from its original temperature of trillions of degrees to the temperature we now measure, only 3 kelvins, that is, 3°C (5°F) above absolute zero. A wide-ranging variety of special telescopes on the ground to observe this background radiation on different spatial scales, following the fantastic success of the Cosmic Background Explorer spacecraft and its continued data reduction, will keep our understanding growing of the 'early Universe' in which the background radiation was set free.

NEW TECHNIQUES AND INSTRUMENTATION

The Mount Wilson telescope, opened in 1917, was the largest in the world until the Palomar 5-m (200-inch) telescope opened, also in California in 1948. With the exception of a Soviet 6-m telescope that never reached the quality of the world's other large telescopes, no larger telescope was built until recently.

But astronomical observing capabilities were improved nevertheless. For many years, most observing was done with photographic film (or, when the photographic emulsion was on glass plates instead of plastic film, photographic 'plates'). Film and plates had efficiencies of only about 1 per cent; that is, only about 1 per cent of the photons of light that hit them made the chemical change that we could detect.

Over the 1980s, various types of electronic detectors were brought to such a point at which they took over from film for astronomical use. The detector of choice at the moment is a CCD, basically a silicon chip (like a computer chip) but one that is light sensitive. CCD stands for 'charge-coupled device,' since the incident photons make a charge on the chip, and the amount of charge at each small picture element ('pixel') is read off by coupling it to an adjacent pixel. A computer controls the readout. CCDs with 2000 pixels in each direction, totalling 4 million pixels in all, are now common in major observatories. Even larger ones are under development, since they do not yet cover the wide fields that film is capable of.

15

But among the important advantages of CCDs over film is the fact that their sensitivity is over 50 per cent; that is, over half the incident photons are detected. Thus CCDs are over 50 times more efficient than film. (In the infrared, just past the red end of the visible spectrum, the gain is even greater, since film is very insensitive there.) Thus a given telescope can record an image in one-fiftieth of the time, roughly, that it took with film. A spectrum of a distant galaxy that used to take an hour to record with film now can be recorded in roughly a minute.

Another development in taking galaxy spectra is the use of fiber optics. These long strands of pure glass carry light by internally reflecting it from side to side with very low losses. The breakthrough was the idea of making a metal plate that is placed at the focus of the telescope with the end of a fiber at the location of each galaxy image. Thus 100 or more galaxies shine on the optical fibers at a time, one galaxy for each fiber. The other ends of the optical fibers can be brought together in a line that is placed at the entrance to the spectrograph, so that all 100 spectra can be obtained simultaneously. Then the same one minute just described brings not one spectrum but 100. Such techniques allow thousands of galaxy spectra to be measured, and thus distances found through Hubble's law for many many objects. The distances, paired with positions on the sky from the direction to the galaxy, give the three-dimensional position in space of the galaxy and allow large-scale maps to be made in reasonable times.

Most of the large telescopes in the past were multi-purpose instruments, with astronomers applying for time in whatever field of research they were most interested in. Some of the new telescopes under construction will be single-purpose devices, though. The huge new Spectroscopic Survey Telescope of Pennsylvania State University and the University of Texas, for example, will be devoted to spectroscopy. It will be made up of 85 1-m mirrors, producing the equivalent of an 8-m telescope, and will be near Fort Davis, Texas. It will not move to track the stars, and will observe only the objects that pass overhead. Another major telescope under construction is for the Sloan Digital Sky survey, being put together by a consortium including the Institute for Advanced Study at Princeton, Princeton University, the University of Chicago, and the Fermi National Accelerator Laboratory. This 2.5-m telescope will use four sets of CCDs next to each other, each looking through a different color filter. Rather than tracking the stars, the telescope will sit stationary while the sky rotates overhead. The CCDs will be read out at the same rate the sky moves. Data from succeeding nights will be put together in a computer to detect faint galaxies and quasars in the field of view, enabling the telescope to go back and take spectra. The telescope will be on Apache Point in the Sacramento Mountains of southern New Mexico. A 4-m Cambridge–Cambridge telescope that the Harvard–Smithsonian Center for Astrophysics in Cambridge, Massachusetts, and the Institute of Astronomy in Cambridge, England, hope to build jointly will be devoted to measuring redshifts of southern-hemisphere galaxies.

NEW TECHNOLOGY TELESCOPES

Over the last decades, the efficiency of optical telescopes has been raised by a factor of 50 or so by the replacement of film by CCDs. But since maximum efficiency is 100 per cent, we are already close to that goal and no additional factor of 50, or even of 10, is available. So astronomers have gone back to the drawing board and designed a new generation of large telescopes.

The first of this new generation of telescopes to be completed is the W. M. Keck Telescope (Fig. 1.10), equivalent to a mirror 10 meters in diameter. Since it is twice the diameter of the Palomar Observatory's 5-m reflecting telescope, it can collect $(10/5)^2$ times more light, and thus image an equivalent object in one-fourth the time. The Keck Telescope was opened in 1993 on Mauna Kea, the 4000 m (13 800 foot) mountain on the island of Hawaii in the state of Hawaii.

The Keck Telescope is made of 36 hexagonal mirrors each 1.8 meters across, rather than being made of one single mirror. This construction allowed it to be made at much lower cost than a single mirror would have, and allowed it to be made with greater curvature so that the focus is relatively closer to the mirror, allowing less steel to be used and a smaller building to be made. The $78 million contribution of the W. M. Keck Foundation through Caltech, with running costs of the telescope to be paid by the University of California, represent a new way of financing telescopes in addition to the breakthrough in design. The Keck Foundation has liked the development of the Keck Telescope so well that they have also paid for a twin telescope, Keck II, now being erected alongside the first telescope. (They have also paid about 1 per cent of the cost for the Keck Northeast Astronomy Consortium, an undergraduate group containing faculty and students from Williams, Wellesley, Wesleyan, Middlebury, Colgate, Vassar, Swarthmore, and Haverford, as part of training of the next generation of astronomers with CCDs and computer workstations.)

The site, Mauna Kea, is also providing an advantage to astronomers. The quality of the image, the 'seeing,' is better there than almost any other astro-

Fig. 1.10. The Keck Telescope on Mauna Kea. (Jay M. Pasachoff)

nomical site, allowing images of point objects like stars to be smaller. The smaller images are more concentrated on individual pixels, and so allow fainter objects to be seen. Thus Mauna Kea is or will be the site of several of the world's largest telescopes (Fig. 1.11). The University of Hawaii gets some of the observing time on each of these telescopes – 10 per cent of the Keck time, for example – since it is the state university of Hawaii, which controls the site.

Alongside the new Keck dome is not only Keck II but also the dome for the 8.2-meter telescope under construction of the Japanese National Astronomical Observatory. This telescope is named Subaru, which means 'Pleiades,' a cluster of stars in Taurus. First light is scheduled for 1999. Also planned for Mauna Kea is an 8-meter telescope of the U.S. National Optical Astronomy Observatories. The Gemini project is to build two such telescopes, one in the northern hemisphere on Mauna Kea, opening in 1998, and one in the southern hemisphere on Cerro Pachón in Chile, opening in 2000. The U.S. government is paying for half the total expense, the British and Canadians are paying another 40 per cent, and the Brazilians, Argentineans, and Chileans are paying the last 10 per cent. These 8-meter telescopes are being made with relatively thin mirrors which will be actively supported across their backs by structures that, controlled by computers, can adjust the shape of the mirrors to keep them accurate. Note in the drawing how any one of these giant telescopes is larger than all the original telescopes on the mountain together in terms of area of mirror exposed to starlight.

Another 8-meter telescope mirror is being made with a different technique, pioneered by Roger Angel at the University of Arizona. Angel spins molten glass in a furnace and allows it to cool in a parabolic shape. He is making, or has made, several telescopes in 3.5 and 6.5-meter size. One of the latter is for the Magellan Telescope, a joint project of the University of Arizona and the Carnegie Institution of Washington, and is to start observations in Chile in 1996. A second 6.5-m mirror is to replace the several mirrors of the Smithsonian Astrophysical Observatory's MMT, formerly called the Multiple Mirror Telescope, and is to start observations on Mt. Hopkins in Arizona in 1996. Angel's first 8.4-meter mirror is to go into the Large Binocular Telescope (only one of whose mirrors is funded so far), formerly called the Columbus Telescope, on Mt. Graham in Arizona. It is a joint project of the University of Arizona and the Arcetri Observatory of Florence, Italy, with additional funding from the Research Corporation, a grant-making organization based in Tucson, and could have first light in 1997.

The largest project of all these is the Very Large Telescopes, a set of four

PASACHOFF

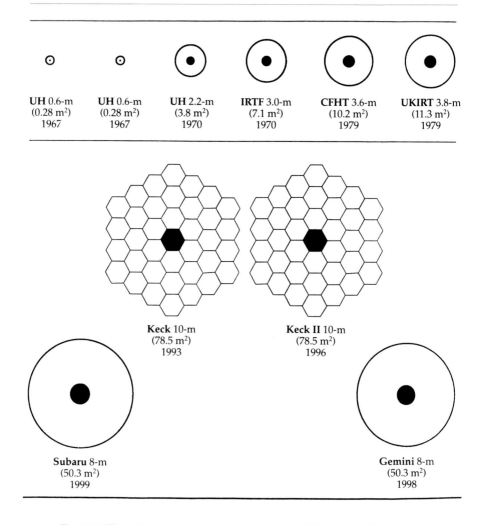

Fig. 1.11. The relative sizes of telescope mirrors on Mauna Kea. The top row shows the telescopes already there when the Keck I telescope was erected. UH stands for University of Hawaii; IRTF stands for Infrared Telescope Facility; CFHT stands for Canada–France–Hawaii Telescope, and UKIRT stands for United Kingdom Infrared Telescope. (Institute for Astronomy, University of Hawaii)

8.2-m telescopes being erected on Cerro Paranal in Chile by the European Southern Observatory, paid for by many European nations. The telescopes will be erected one by one starting in about 1995, and should be usable either singly or together.

Fig. 1.12. The Hubble Space Telescope, photographed from a space shuttle soon after its launch with an IMAX camera. (NASA, Smithsonian Institution/ Lockheed Corporation)

So while 1990 found the world with a 6-m, a 5-m, and smaller telescopes, the new millenium will boast two 10-m telescopes and at least eight 8-m telescopes for general purposes. Surely the amount of light – and the infrared radiation – collected by these telescopes will bring breakthroughs for the hundreds of dedicated astronomers using them.

21

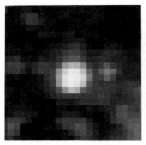

Fig. 1.13. Comparison of (left) ground-based observations of a field of stars with Hubble Space Telescope images taken (center) before and (right) after the 1993 replacement of the first Wide Field and Planetary Camera with WFPC-2. (Space Telescope Science Institute and NASA for Hubble images; Georges Meylan/European Southern Observatory for the ground-based image.)

TELESCOPES IN SPACE

The Hubble Space Telescope (Fig. 1.12), launched in 1990 and brought to full usability in 1993, brings its important set of unique abilities to astronomical service. For the Key Project on the cosmic distance scale, the results discussed above for Cepheid variables in the spiral galaxy M81 used the fact that even in its pre-repair state, the star images had a 0.1 arc sec central core containing about 15 per cent of the light, that is smaller in size than we can obtain from telescopes on Earth, usually 1 arc sec and never better than 0.3 arc sec. Surprisingly, it further turned out that among the important advantages of the Hubble Space Telescope for this study was the fact that its position above the Earth's atmosphere allowed observations to be scheduled at optimum times for the variable stars, without limitations from daylight, moonlight, or bad weather. Now that over 80 per cent of the light goes into the central 0.1 arc sec (Fig. 1.13), we expect the cosmic distance scale to be determined to 10 per cent within a few years (Fig. 1.14).

The next in NASA's Great Observatories is to be the Advanced X-ray Astrophysics Facility, AXAF. It is managed at NASA's Marshall Space Flight Center in Alabama and built by a team led by TRW, Inc. The AXAF Science Center, with responsibility for helping scientists to use AXAF and for designing the software to analyze the data, will be located at the Smithsonian Astrophysical Observatory in Cambridge, Mass. AXAF was reconfigured into two missions for financial reasons. AXAF-I ('I' for imaging), will use four nested cylindrical mirror-pairs to make high-resolution X-ray images and perform grating spectroscopy. (Since each of the cylinders is viewed almost

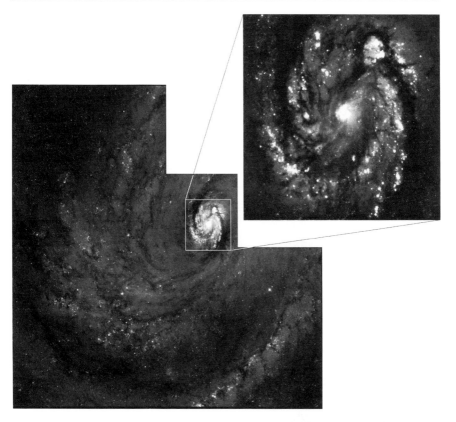

Fig. 1.14. M100, a spiral galaxy in the Virgo Cluster of Galaxies, imaged with the repaired Hubble Space Telescope. The new observations should give a distance to the Virgo Cluster with gratly improved accuracy and so provide an important stepping stone to improved distances to all the more distant galaxies. The Wide Field Planetary Camera 2 on the Hubble Space Telescope, installed by the astronauts in 1993, consists of three square charge-coupled devices that form an L, the 'wide-field camera,' with a 'planetary camera' part that appears at the upper right of the L. Since the planetary camera (so named because it will often be used for imaging Jupiter and other planets) has twice the resolution of the wide-field camera, but the same number of pixels in its charge-coupled device as each of the three elements mosaicked in the wide-field camera, its image appears smaller. Here the central spiral of M100 imaged by the planetary camera is enlarged at upper right, showing off the full resolution of the planetary camera. (Space Telescope Science Institute).

end on, the use of concentric cylinders increases the area that can focus the incoming radiation.) It is to be launched in 1998. AXAF-S ('S' for spectroscopy), was to fly the calorimeter with a less expensive set of mirrors to do broad-band spectroscopy. Unfortunately, this second mission was cancelled in 1993. (One of its descoped instruments may eventually be launched on a Japanese spacecraft.) X-ray observations made in years past have shown not only nearby objects but also an all-sky X-ray background from far away as well as many distant galaxies and quasars, so the capabilities of AXAF should join with those of optical and infrared telescopes to tell us about the farthest things in the Universe.

CHAPTER TWO

·

THE COSMIC BACKGROUND RADIATION

·

ED CHENG

Looking up at the clear night sky, it is hard to avoid wondering about the many objects that we can see. It is simple to recognize with the naked eye that there are planets, countless stars, and the band of light from the disk of our own Galaxy, the Milky Way. With the help of binoculars or a small tele-scope, the complexity of the scene increases dramatically, and it becomes apparent that the glow of the Milky Way is the light from many faint stars. We also start to notice that there are numerous faint and fuzzy objects which are the nearby galaxies and the star-forming regions in our own Galaxy. Probing with more and more sophisticated instruments, the level of detail and structure that can be resolved using visible light increases until the light becomes so exceedingly faint that even the best detectors on the largest tele-scopes see only darkness. This is the regime of the farthest objects in the Universe.

Before discussing these objects in any detail, I would like to take a brief moment and address the question of how we can possibly know about things so remote in both distance and experience. After all, we invent and test the physical sciences here on Earth by making experiments, interacting with the world around us, and creating a system of beliefs (theories) that ties all these experiments together into a consistent and testable story. In astronomy and cosmology, we use this basic knowledge of how the world works to explain phenomena that are so far away that, as we shall see later, their light has taken the age of the Universe to travel to us. We play the role of passive observers who can look with any amount of cleverness, but touching and interaction are strictly prohibited. What exists are the observations and the theories that attempt to explain them. If we do not have or cannot make a particular observation that we are curious about, we must invent a devious method which indirectly answers the question.

The process works something like this. First, we take our local experience and look at nearby objects to try to understand them. If necessary, we can make new assumptions and theories, but these must be consistent with all known observations before they gain general acceptance. After the nearby objects become familiar, we take the next step and try to apply what we have learned to objects that are farther away, and so on. However, the farther we go, the more we have to apply indirect reasoning to explain what we see, and the less certain we are of a particular explanation or quantity. It is crucial that we test each step in reasoning against observations. In particular, a useful theory must make predictions which can then be used to verify or falsify itself. It is remarkable that we have a plausible story that fits most of the observations and takes us all the way back to an event that may be interpreted as the birth of our Universe. Knowing the age of the Universe, even if it is only to an accuracy of a factor of two, is an extraordinary accomplishment.

Going back to the original question then, of how we know about these objects, the answer is that we don't really know in the sense of my being able to prove to you by showing you something concrete. I can, however, tell you an interesting and pretty self-consistent story that fits most, if not all, of the observations. If someone comes up with another story that explains the observations equally well, and is not full of untestable assumptions, then it, too, is an equally interesting story to tell. In any case, the theory gives us a framework in which to describe and explore the Universe.

In the discussion that follows, it is important to keep in mind that even some of the most basic parameters we will be discussing, such as distance or age, are not things that we measure directly, but are deduced and calculated based on theory. We do not measure a distance in the same way as we would use a measuring stick, nor can we measure age as we would use carbon-14 dating. I will try to make the distinction between those things that we actually see, and the derived, 'model dependent', quantities specified by theory.

HOW FAR IS FAR?

Before we can meaningfully discuss the farthest objects, we should examine exactly what we mean by 'far.' Because the distances are so large (indeed some are 'astronomical'!), we need a bookkeeping system and a way of relating these scales to something more normal.

The bookkeeping system used here is scientific notation, where 10^6 means a 1 with 6 zeroes after it (otherwise known as one million), 10^0 is 1.0, and 10^{-2} is 0.01. Thus 10^8 is 10^2 (100) times bigger than 10^6 (the exponents add when numbers are multiplied together). Even though 10^8 or 10^6 may be too big to grasp, it is easier to visualize that 10^8 is 100 times bigger than 10^6.

When large factors appear, it is good to keep some common references in mind. Factors of 10 and 100 are relatively commonplace and are therefore quite intuitive. A respectable mountain is about a factor of 1000 taller than a human being, and a football field is about a factor of 10 000 bigger than a dime and 100 000 times bigger than a gnat. Table 2.1 uses these factors to span the size of the Universe. For the astronomical objects, each step illustrates the next larger structure in the Universe that we observe. Notice that everything is rounded to the nearest factor of 10, so while a human is more like 1.5 or 2 meters, 1 meter is a close enough approximation. Ignore the light-travel-time column for now.

Most of the entries are self explanatory, and I'll start with the solar system size and then move up. I take the solar system size to be the diameter of Pluto's orbit. If we imagine that the solar system is a human sized object, then the next kind of distance that we come across, the distance to the near-

Table 2.1. *Sizes and distances*

	Distance		Factor
	(meters)	(light-travel-time)	
Observable Universe	10^{26}	10^{10} years	100
Diameter of 'voids'	10^{24}	10^8 years	10
Cluster diameter	10^{23}	10^7 years	10
Distance to Andromeda galaxy	10^{22}	10^6 years	10
Galaxy diameter	10^{21}	10^5 years	100 000
Distance to α Centauri	10^{16}	1 year	1000
Solar system size	10^{13}	10 hours	100
Earth orbit diameter	10^{11}	10 minutes	100
Sun diameter	10^9	10 seconds	10
Moon orbit diameter	10^9	1 second	10
Earth diameter	10^7	0.1 seconds	10
NY to LA distance	10^6	0.01 seconds	10 000
Skyscraper height/football field	10^2	1 microsecond	100
Human	10^0	10 nanosecond	1000
Gnat	10^{-3}		1000
Cells	10^{-6}		10 000
Atoms	10^{-10}		100 000
Nuclei	10^{-15}		

est star, is about the height of a mountain. There are many (approximately 1 trillion or 10^{12}) of these stars forming our Galaxy, which is 100 000 times bigger than the distance to the nearest star. Galaxies are also social creatures, and tend to form clusters of about a hundred or so individuals. The size of these clusters is approximately 100 times the size of a galaxy. It turns out that the distance between our Galaxy and the nearest large galaxy (Andromeda) is about the same as the size of one of these clusters.

The next level of structure is a relatively recent discovery. It appears that clusters of galaxies are not uniformly distributed in space, but they form surfaces that surround relatively empty regions of space called 'voids'. This means that the distribution of galaxies and clusters look more like a sponge, where there are empty spaces (the voids) delineated and surrounded by relatively thin material (the galaxies). This is not immediately apparent when looking at the sky, since we need to know how far away an object is before we can map it in three dimensions. The voids are so large it takes painstaking effort over many years to gather enough data to form a coherent picture. Finally, when we get to a scale which is about 100 times bigger than the voids, we start to reach the limits of the observable Universe. By 'observable', I mean being limited by the nature of the Universe itself, as we shall see later, rather than by the crudeness of our instruments.

Two curiosities to note are that the Universe is larger than normal sized objects (like people) by a much bigger factor than the smallest subatomic particles are smaller. While this probably has no fundamental significance, it is an interesting point to keep in mind and underscores our position in the physical Universe! Another curiosity is that the Universe, starting from the smallest subatomic particles on up, is essentially empty space.

HOW DO WE MEASURE DISTANCES?

Now that I have asserted that there are these phenomenal sizes and distances, it is time to discuss how we measure these distances. For nearby objects, we can use the Earth's orbit as a baseline to measure parallaxes, the apparent change in direction to an object as the observer changes position. This is similar to the triangulation used by surveyors on the Earth and the effect is the same as measuring the shift in apparent position of a nearby object when looking with alternate eyes closed. Using the Earth's orbit instead of the separation between our eyes, this method works with distances up to about 10^{18} meters, or about 100 times the distance to the nearest star. To get beyond this scale, we need more complicated methods which use the measured properties of astronomical objects as 'standard candles' or 'standard rulers'. The candles are the apparent brightness of a 'standard' object. We infer the distance by noting that the brightness diminishes as the reciprocal of the distance squared. The rulers are the angular size of another set of 'standard' objects on the sky, which varies as the reciprocal of the distance. Because the characteristics of these standard objects are not precisely understood, and because these objects are not all the same, these estimates of distance are good to roughly a factor of two.

On the scale size of galaxies, clusters of galaxies and larger, Edwin Hubble noticed in the 1920s a remarkable linear relationship between the distance of an object and its 'redshift', an unambiguous measurable quantity, and that this relationship is true no matter what direction in the sky is being observed. Over the years, we have refined the particulars of his measurements, but his observations remain as one of the cornerstones of modern cosmology. I'll digress on the redshift before returning to why the Hubble relationship is so important.

An object is said to be redshifted when the observed light from the object appears to be shifted towards the red part of the spectrum. In other words, it looks 'redder' than it should. This is usually the result of the emitting object's motion relative to the observer. The more general physical effect is the Doppler shift, which is an easily observed effect in almost all wave phenomena (such as light and sound). A common example is the sound of a jet engine. As the jet approaches, the pitch of the engine sounds higher than when the jet is at rest, and as the jet recedes the reverse is true, that is, the pitch is lower. We can visualize this change in pitch using the following analogy. Imagine two ducks sitting in the middle of a big lake. The first is a lobotomized duck that is free from the cares of the world and simply bobs up and down in the middle of the lake. It will generate waves in the water that look like perfect circles with the duck at their center. Now imagine an excessively libidinous duck which sees its mate at the far end of the lake. In addition to bobbing, this duck will move at great speed to meet its lover. The wave crests are now closer together in front of the duck and farther separated behind when compared to the first case (see Fig. 2.1). For sound, waves with closer crest spacings correspond to higher pitches, and for light, they correspond to the bluer part of the spectrum. In cosmology, the variable z represents the redshift when we define it as

$$z = \frac{\lambda - \lambda_0}{\lambda_0}$$

where λ (lambda) is the measured wavelength of the observed light (distance

Fig. 2.1. Two ducks illustrating the Doppler shift (see text).
(S. Shore)

between the crests) and λ_0 is the wavelength of the emitted light. The variable z is equal to 0 if the observer and emitter are at rest relative to each other. For the analogy with the rapidly moving duck, the measured wavelength is the spacing of the crests in the direction of the observer, and the emitted wavelength is what the bobbing motion would have generated if the duck remains stationary. One way that we know the wavelength of the emitted light from far-away objects is by identifying spectral lines. These are specific colors where an object emits or absorbs light strongly. By identifying patterns of these lines with ground-based measurements, we can deduce the emitted wavelength of a set of observed spectral lines. Here is a good example of where we depend heavily on very 'local' measurements to deduce the characteristics of distant phenomena. The assumption made is that the wavelengths measured in our laboratories are the same as those emitted by the distant objects. The fact that large numbers of these spectral lines are identifiable and appear to shift in the same way (giving a consistent redshift measurement) encourages us to think that the physics occurring 'out there' is not so different from our local experience.

Going back to the Hubble relationship, the fact that the light from distant galaxies is redshifted implies that they are moving away from us (we are looking at the tail end of the libidinous duck), and the fact that the redshift increases with distance implies that the farther objects are moving away or receding at faster speeds. To be specific, the 'Hubble Law' states that an object moves away from us at a speed of H_0 (a constant) times the object's distance. The value of H_0 is somewhere between 0.015 and 0.03 meters per second per light-year. Thus, at a distance of one light-year (the distance that light travels in a year, or roughly the distance to the nearest star), this velocity is around 3 centimeters per second, a rather small velocity. It is only when we consider extremely large distances that this effect becomes significant.

A question that immediately comes to mind is why we seem to be at the center of this motion, as if we were cursed by a case of cosmic bad breath. Throughout the ages, cosmologies and world views have been invented that place humanity or our home at the center of the Universe, and these have invariably been wrong. Fortunately, Einstein's theory provides a framework which avoids this pitfall. The relevant insight is that if the Universe were uniformly expanding, then every observer would see objects moving away at a speed proportional to distance, and there really need not be a center. A way to visualize this is to imagine that the Universe is two-dimensional, and we all live on the surface of a balloon. Assume that we can only move on the surface of this balloon, and that nothing outside the surface has any meaning or analogue in our daily experience. Blow the balloon up just a little, and place three marks, two of them close together (call these A and B) and one farther away (C). As the balloon gets blown up, A and C will move apart more quickly than A and B, and each point will see the others moving at a rate which is proportional to distance. Thus, not only are we not at the center of the expansion, but this world view suggests that there is no center (in the Universe which contains only the surface of the balloon). The fact that the balloon has a center is apparent only when we become three-dimensional creatures again and can describe the balloon in a space which is more than just the surface and has 'volume.'

The Hubble Law is relatively accurate for distances less than that to the nearest voids. However, when we start to talk about distances comparable to

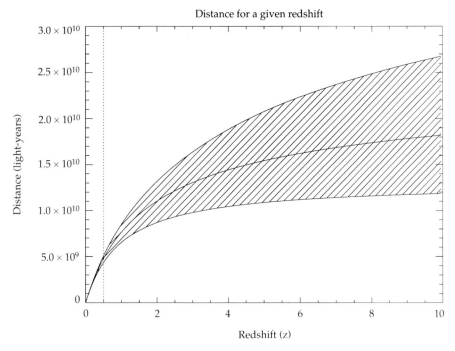

Fig. 2.2. Redshift versus distance for various values of Ω, a parameter that determines whether the universe is open and infinite (top curve), flat (middle curve), or closed and finite (lowest curve).

the size of the Universe, the relationship between the redshift and physical distance is no longer simple. The Einsteinian framework is still valid, but we do not know the total mass in the Universe, a crucial parameter, well enough. This parameter determines how the expansion rate changes as the Universe gets older, and thus the value of the Hubble 'constant' at very early times. However, for any assumed parameter values, the model allows us to deduce a physical distance corresponding to a redshift. Because of this relationship, and because it is generally not too useful to speak of the physical distance to the farthest objects, the redshift is a useful surrogate in place of the distance. Fig. 2.2 gives a rough conversion from redshift to distance (measured in light-travel-time) for a variety of parameters. Indeed, it may be possible to determine some of these basic parameters if we could measure the distances to enough far-away objects to establish the detailed shape of this curve. In practice, many effects conspire to make this difficult. The usual problem is that the far-away objects are sufficiently different from the ones we can study nearby that their characteristics are somewhat ambiguous.

HOW DO DISTANCE AND AGE RELATE?

It turns out that as we see farther away, we are looking at the Universe at an earlier time. This is a necessary consequence of light traveling at a finite speed (300 000 kilometers per second). Thus, if an object were exactly 300 000 kilometers away, the light that is currently reaching our eyes must have been emitted by the object exactly one second ago. I have put in Table 2.1 the light-travel-time for each of the distances and sizes.

It may be good to clarify at this point that being able to see objects as they were at an earlier time than now does not mean that we can see what we ourselves were like at some time in the past. We only see far away objects as they were at some point in the past that corresponds to their light-travel-time distance. When we look at the Andromeda galaxy, we see it as it was approximately 2 million years ago. There is no way to see what it looks like today since that light has not reached us yet. Nor can we see what Andromeda looked like at a time earlier than 2 million years ago because that light has passed the Earth already. There is only one Andromeda and we can only see it as it was 2 million years ago. However, looking at objects at various large distances does give us an idea of what the Universe, on average, looked like at various times in the past. These observations are the only hints we have as to how the Universe evolved with time, and what objects were typical of a certain age.

If we take the Hubble Law and trace it backwards in time, then we come to the conclusion that, at some time in the past, the entire contents of the Universe must have converged onto a single point. The time since this event is the age of the Universe, and is between 10 and 20 billion years.

Accepting this notion that the Universe has lived for a finite period of time, we must also accept the fact that we cannot 'see' objects which are infinitely far away. In order to see something, light has to propagate from that object to the observer. If the Universe has only been in existence for a finite time, then the distance that light can travel during that time must also be finite. For example, if the Universe were exactly 15 billion years old, light from places which are over 15 billion light-years away simply has not had time to travel to us yet. Not only do we not see these regions of space, but because no physical law causes effects which travel faster than the speed of light, our observable Universe must have evolved completely independently of such regions. The distance to which an observer can see, for a given age of the Universe, is called the observer's 'event horizon.'

WHAT PHYSICAL CONDITIONS CHARACTERIZE THE VERY YOUNG UNIVERSE?

In the model of the Universe we have been discussing, there was a time in the past when the Universe was very small. This small volume must have held most of the energy in the Universe today. By energy, I mean the total amount of energy in radiation (given by the frequency of the radiation times Planck's constant times the number of photons) and matter (given by the mass times the speed of light squared). Thus, when the Universe was smaller, the amount of energy in a given volume (energy density) must have been higher. This intuitive argument is confirmed by calculations which take into account many details which are not central to our discussion here.

An equivalent and convenient way to describe energy or energy density is to use the concept of temperature. Loosely speaking, when we say that the temperature of an object is increasing, what we mean is that its thermal energy is increasing. Similarly, when we speak of the temperature of a volume of space increasing, we are saying that the total energy density in that volume, including the matter and photons (light) within it, is increasing. Thus, as we go back in time and the Universe gets smaller, we can say that its temperature increases. This is not unlike packing the same number of people into smaller and smaller rooms. The calculations show that this characteristic temperature of the Universe goes inversely as the scale size of the Universe (temperature doubles when the size is halved). Consequently, there must have been a time in the early Universe when everything in it was very hot. These are the conditions which led to the term 'Hot Big Bang' for this model, the 'hot' from the temperature, and the 'bang' from the expansion of the Universe from the initial point.

We have made an implicit assumption that the temperatures of large portions of the Universe, at any given time, are the same. This condition is called thermal equilibrium, and is a natural state that is reached when the time scale for changing the temperature of any region of space (by some external influence) is much longer than the time it takes for the constituents of the region to interact and 'spread' the energy density around. For example, a well-insulated portable cooler with a newly inserted pot of boiling water is clearly not in thermal equilibrium. After many hours, the water will cool off and the cooler will warm up, so that the inside of the cooler is getting close to thermal equilibrium. However, it is still warmer than the outside so that eventually, the water, the cooler, and the air surrounding the cooler will come to the same temperature. There are many wonderful calculable features of a system which is in (or close to) thermal equilibrium. One of these is the Planck blackbody radiation.

In the early 1900s Max Planck made an important breakthrough calculation, based on relatively simple theoretical considerations, that when a system is in thermal equilibrium, it should emit electromagnetic radiation (light, radio waves, X-rays, etc.) in a very specific way. He derived a formula that, for any given temperature, predicts the amount of radiation at any given wavelength, with the peak wavelength of the radiation increasing inversely as the temperature (double the temperature and the wavelength is halved). This 'Planck Law' uses a temperature parameter based on the kelvin scale, where 0 kelvin is the temperature which has no emission (infinite wavelength), 273.16 kelvin is the freezing point of water, and 373.16 is the boiling point of water. One degree kelvin is the same size as one degree Celsius, but the zero point is different (the Celsius scale is defined to be zero at the freezing point of water). The Planck Law describes the phenomena we see when heating up objects in an intense flame. The object will first look 'normal' to our eyes, then start glowing a dull red, and subsequently (if the flame is hot enough) become bright orange and then white. At a normal room temperature of 30° Celsius, all objects have a peak emission in the infrared at about 10 microns. Our normal vision sees between 0.4 and 0.7 microns (blue and deep red, respectively), so this emission is not normally visible to us. If we had 10 micron infrared eyes, we would see this radiation everywhere and the world would always look bright (everything glows!).

Why is this called 'blackbody' radiation? In deriving the Planck spectrum, we make the assumption that the surface absorbs all the light which hits it (and thus would appear 'black' to our eyes). As a practical matter, such a material is not possible to find. It turns out that if we consider the radiation inside a box whose sides are at the same temperature, it does not matter much what the walls are made of, the reason being if the light is not absorbed on the first bounce, it has many more opportunities on subsequent bounces. The radiation inside such a box is in thermal equilibrium with the walls, and provides a good example of blackbody radiation.

A very dramatic illustration of this effect is provided by the furnace at the local glass factory. This furnace is exactly the box we need (ignore the small access hole, which is not important). Looking inside the furnace, one sees a uniform bright orange glow which is the Planck blackbody radiation for the temperature of the furnace. A worker typically inserts a boule of glass into the furnace with a long steel pole. As the boule goes in past the hole, it disappears! What happens is that the contrast is going down, and the glass and pole are glowing at the same color as the blackbody radiation of the furnace. The worker will keep the glass inside the oven for a little while to make sure it is in equilibrium with the furnace, and then withdraw it for shaping. As the glass comes out, it becomes visible again and will glow with the same color as the furnace. As it cools down, the glow becomes fainter and shifts towards the red end of the spectrum, until finally it appears to stop glowing (at least to our eyes). This furnace is a useful analogy for the world view we are discussing, except that the entire Universe is the furnace!

Putting this all together, then, the decrease of energy density or temperature as the Universe expands also implies that the blackbody radiation increases in wavelength. Detailed calculations show that if the initial spectrum is that of a Planck blackbody, cosmological expansion will retain the shape of this blackbody spectrum, and will make the temperature appear to cool off (the temperature varies inversely as the size). Thus, if at a certain time we created a blackbody spectrum of 1000 kelvins, after expanding 10 times, this spectrum will be at 100 kelvins.

In addition to the temperature, we can also make predictions about the state of the constituents of the early Universe. At any given temperature,

Table 2.2. *Temperatures and phenomena*

Temperature (kelvins*)	Phenomenon	Peak Planck spectrum (microns)
0	Absolute zero	(infinite)
4	Liquid Helium boils	690
77	Liquid nitrogen boils	38
220	Typical temperature for winter at the Earth's poles	13
273	Water ice melts	11
300	Room temperature	10
373	Water boils	7.8
3000	Hydrogen atom splits into an electron and a proton.	
10^4	Temperature of the surface of the Sun.	
10^{10}	Electrons form spontaneously, along with antiparticles, created from electromagnetic energy.	
10^{13}	Protons and neutrons form spontaneously.	

Note: *One degree on the kelvin scale is the same as one degree on the Celsius scale. The freezing point of water is 273.16 kelvin. Zero kelvin is absolute zero, a temperature that corresponds to no thermal energy. Absolute zero cannot be reached because heat is always leaking in from the surroundings, but it can be approached. For example, it is the temperature that would be approached if the energy density of the Universe were held constant and the Universe were to expand to be infinitely large. In low-temperature physics laboratories, temperatures of around 0.001 kelvin are routinely attainable.

there are natural phenomena that are more likely or less likely. For instance, we do not expect water to boil spontaneously on a cold winter day. The reason is that the thermal energy available at such temperatures (from the air) is not sufficient to stimulate the water molecules to move away from each other and form a gas. We can extend this argument from commonly occurring phenomena (at commonly occurring temperatures) to more exotic phenomena such as the dissociation, or spontaneous breaking apart, of fundamental particles. While we have very little experience with large scale exotic phenomena which involve immense temperatures and energies, it is possible in high-energy particle physics research to perform studies of relatively few fundamental particles at high characteristic energies (and equivalent temperatures). From these data, we can deduce the likely behavior when the entire Universe was at the same characteristic temperature. Table 2.2 gives an idea of the physical processes which become important at various characteristic temperatures.

What happens when the Universe was very young and very very small? This naive model says that there was a time when the Universe was infinitely small and temperature was infinitely hot. This is the so-called 'singularity' in the Big Bang model and is not really very satisfying since such phenomena are not possible to compute and thus not fruitful to discuss. Such bizarre results usually point to a lack of understanding of what is going on. Exactly what happens under these conditions is unclear because we do not know the kind of physics which becomes important at the high temperatures involved (this is an area of active research). However, as we shall see, there are already important consequences of the parts of this world view that we are able to probe with currently understood physics.

WHAT DO WE EXPECT TO SEE?

Going back in time from the present, a reasonable first milestone to look for is the formation of the stars and galaxies we now observe. Before this, the Universe will have been considerably warmer, and we can imagine looking for the epoch at which atoms are formed.

Let us consider the time when the Universe was hot enough to keep hydrogen atoms separated into their constituent parts, electrons and protons. The radiation and the charged, relatively low mass, free electrons will interact strongly, bumping into each other frequently and changing direction randomly (scattering). In turn, the negatively charged electrons will 'drag' the positively charged protons along. These interactions are sufficiently strong that the total energy will be divided evenly among the protons, electrons and radiation, and a thermal equilibrium state will be reached. Thus, the situation is like a cosmic 'soup', with the matter being in equilibrium with the blackbody radiation appropriate for the temperature of the Universe at that time. Trying to look at radiation that was generated before this time is somewhat like looking through a dense fog. It is hard to see any detail at all behind the fog, but not hard to notice that there is light there (the fog changes the direction of the light in a random way, much like the electrons).

Immediately after the Universe cooled sufficiently to allow the electrons and protons to combine into hydrogen atoms, the scattering decreased dramatically since there were no more free electrons, and photons were free to propagate unimpeded. This is the period of 'decoupling' (of radiation from matter) and is important because the spatial distribution of the photons from this era is determined by the distribution of matter and radiation at this time. When we look at these photons, it is like looking at the outside surface of the fog. We can see the structure on the surface of the fog, but not what is behind it. It is for this reason that we sometimes speak of observing photons from the 'surface of last scattering' for the radiation. Because the decoupling era occurred at a particular redshift, this 'surface' looks like a spherical shell, centered on ourselves, and located at the distance corresponding to the redshift of decoupling.

We can estimate when this occurred by looking at the characteristic temperature of the period of decoupling. From Table 2.2, this occurs at a temperature of about 3000 K. As we shall see below, the current temperature of the Universe is about 3 K, leading to a redshift of about 1000 for decoupling. Since this is a relatively large redshift, deriving the age in years is, as pointed out earlier, model dependent. However, a redshift of 1000 corresponds to a time when the Universe was roughly 300 000 years old. This implies that if the Universe were currently considered to have the age of a human adult, the era of decoupling occurred only several hours after conception! Subsequent to decoupling, the radiation propagates relatively unimpeded until it hits the Earth, except for a gradual cooling off caused by the expansion of the Universe.

We can arrive at one other general conclusion about the origin of the brightness variations at the surface of last scattering. Recall that, at any given time in the history of the Universe, a particular observer will see an event horizon. This is the distance beyond which light has not had sufficient time (since the Big Bang) to reach the observer. Since the Universe was only 300 000 years old at the surface of last scattering, the distance from any observer to the horizon was only 300 000 light-years (roughly the size of a galaxy today). It is an important distance because regions of space which are separated by more than this distance would not have had a chance to communicate (and therefore affect each other) for all times in the past. We can calculate the angle on the sky which corresponds to the horizon distance for an observer at the time of the surface of last scattering. This angle turns out to be about 0.5 to 1.5 degrees on the sky today, depending again on the total mass in the Universe. Thus, for brightness variations on scales smaller than about 1 degree, we would expect the physical processes occurring at the surface of last scattering to take effect and modify the brightness. For variations on scales larger than 1 degree, however, there can be no process at the surface of last scattering which affects the brightness because all regions which are separated by more than 1 degree do not (yet) know about the existence of each other. We would therefore expect both the brightness and mass distribution variations on these large scales to represent a very fundamental property of the Universe. Explaining the characteristics of brightness variations on these large scales is one of the successes of the 'inflation' scenario of the early Universe, which predicts that these variations were imprinted by processes occurring in the Universe when it was only 10^{-35} seconds old, but that is another story.

This notion of the 'scale size' of variations may be a little strange, but it is frequently a convenient way of categorizing physical effects. By variations on scales of 1 degree, I mean variations that, loosely speaking, wiggle up and down once every degree. Variations on larger scales are those which complete one wiggle only after a larger distance. Similarly, variations on smaller scales are those which wiggle up and down more often than once every degree. The observed variations in the sky are typically the sum of variations on many scales.

Before decoupling, the radiation scattering is so strong that we lost any 'images' of what may have come before, but what about the spectrum or color of this radiation? Going back to the fog analogy again, we can certainly see whether there is green or red light emanating from inside the fog. In other words, scattering changed the direction of the light, but not its color or energy distribution. What, then, determines this color? We have already discussed how the Planck radiation spectrum can arise from a very hot initial period in the Universe. In order to change this spectrum, some physical process must be able to absorb and create light with a different spectrum. It turns out that when the Universe was hotter than 10^6 kelvins, it is very implausible to have such processes because the Universe was so hot and dense that we are assured of thermal equilibrium. This temperature corresponds to a time when the Universe was at a redshift of about 10^6 or when it was about 6 months old. To use the human age analogy, this corresponds to 0.05 seconds after conception!

To summarize, then, there are at least two important aspects of the Planck spectrum from the early Universe. The first is the spectrum (color), which probes the Universe back to a redshift of 10^6. This can tell us about the physical processes which took place since that time, or at least the energy scale involved in these processes. We can also look at the spatial uniformity (isotropy) or 'image' of the radiation which shows us variations in the distribution of matter near the period of decoupling at redshift 1000, providing a starting condition for creating the galaxies, clusters and voids which are observed today. Because the peak brightness of the radiation today is near the microwave region of the electromagnetic spectrum, it is also called the Cosmic Microwave Background Radiation or CMBR. The characteristics of the CMBR provide a unique tool for studying the farthest objects in the Universe.

WHAT DO WE SEE?

In the early 1960s two groups of researchers (one at Bell Laboratories and the other at Princeton University) reported that there is an unexplained radio emission from the sky that corresponds to a temperature of about 3 K. The Princeton group proposed the Big Bang scenario as an explanation for the signal. Subsequent observations have confirmed the blackbody nature of this radiation, culminating in the recent measurement by NASA's Cosmic Background Explorer (COBE) satellite that it indeed precisely follows a blackbody spectrum over a large range of wavelengths.

The COBE was NASA's first satellite primarily motivated by the study of phenomena related to the origins of the Universe. It carried three complementary instruments that addressed these fundamental questions:

(1) what is the spectrum (brightness at different wavelengths) of the CMBR?,

(2) what are the brightness variations of the CMBR in different directions?, and

(3) what does the average light from the first galaxies look like?

The instruments are called the Far-Infrared Absolute Spectrophotometer (FIRAS), the Differential Microwave Radiometers (DMR), and the Diffuse

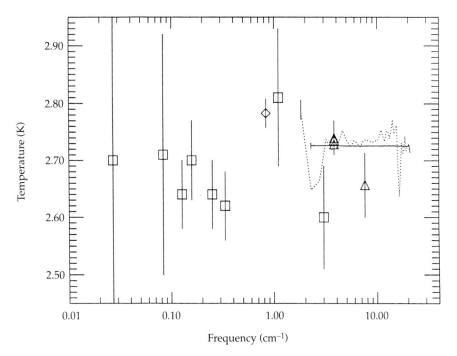

Fig. 2.3. Effective temperatures measured for the CMBR. The energy measured at the various wavelengths has been converted to an effective temperature of blackbody radiation. If the CMBR were a blackbody, then the effective
temperature would be the same for all wavelengths.
(COBE Science Team/NASA)

Infrared Background Experiment (DIRBE), respectively. The COBE was
launched on November 18, 1989, and has performed exceptionally well. The
FIRAS and DIRBE had a lifetime of about 10 months, limited by the amount
of liquid helium cryogen that was carried on board the spacecraft. The DMR
(and part of DIRBE) were turned off in 1993. We will not be discussing the
DIRBE results in this article because it relates to more recent events in the history of the Universe.

Figure 2.3 shows a graph of most of the spectrum measurements from
many experiments, expressed as an effective temperature. What is measured
is a brightness at a particular wavelength. Using the Planck law, this brightness is converted to the temperature that would make a Planck spectrum
have the same brightness at the same wavelength. To show that the CMBR is
a blackbody spectrum, the measurements at multiple wavelengths all need
to indicate the same effective temperature. The reason for concentrating on
this particular wavelength region is because this is where the radiation has

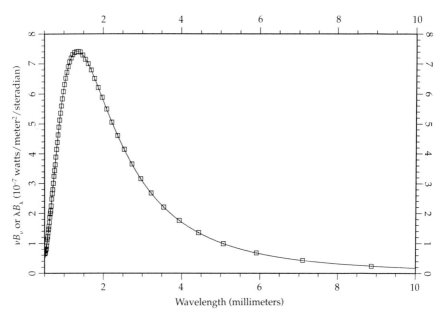

Fig. 2.4. Brightness spectrum of the CMBR. The measurements are shown as boxes. The smooth curve is the Planck blackbody curve. The data points do not deviate from the theoretical curve by more than 0.03% and this amount of deviation is consistent with the measurement error.
(COBE Science Team/NASA)

most of its energy. Figure 2.4 shows the brightness distribution of the Planck spectrum for the CMBR. The boxes indicate the COBE/FIRAS measurements, and the solid line is the best fit Planck curve. The current level of precision for these measurements indicates no measurable deviation from a Planck spectrum to a level of 0.03% of the peak brightness.

To put this measurement in the context of the early Universe, we need to consider the dominant form of energy in the Universe since it is this energy that determines the amount of gravity and thus the expansion rate of the Universe. The average number density of CMBR photons in the Universe today is about 400 per cubic centimeter. Compared to the average amount of matter in the Universe today, which is about 1 nucleon (meaning proton or neutron) per cubic meter, it is 10^9 times more dense. Averaged over the entire Universe, this number of CMBR photons completely overwhelms any other possible source of photons, including star light. In terms of total energy, however, the CMBR today is approximately 0.02% of the total energy in matter today because the CMBR photons have such low energy. This is what we mean when we say we live in a 'matter dominated' Universe. During earlier times when the Universe was smaller, the number of nucleons was still roughly the same as today, and both photon and nucleon number densities increase as the volume of the Universe decreases, with one significant difference. The energy in matter is simply related to the mass (it is roughly the mass times the speed of light squared) and does not change significantly with the size of the Universe, but the energy in the photons increases linearly with their temperature, which we know from the previous section increases inversely as the linear size of the Universe. Thus, at a redshift of z, the linear size of the Universe was smaller by a factor of $1+z$, the temperature of the CMBR was higher by a factor of $1+z$, and the ratio of radiation to matter energy density was roughly $5000/(1+z)$. At a redshift of roughly 5000, the amounts of energy in radiation and matter were about equal, and at earlier times the radiation energy dominated.

Since the CMBR radiation is the dominant form of energy in the Universe at redshifts greater than 5000, it is important to know how much of this energy can be modified by other physical processes which may have occurred. In particular, it is important to know if the Universe was in thermal equilibrium during this time. Thermal equilibrium is a fundamentally simple situation and allows for powerful, sweeping statements without detailed knowledge of the underlying physical processes. In order for the spectrum of the CMBR to be changed from the initial Planckian form which characterizes thermal equilibrium, some process must either create photons of a different spectral distribution, or modify the existing CMBR photons through an appropriate mechanism that changes their energy (color). Given the accuracy of the COBE/FIRAS measurement of the spectrum today, it is possible to place a limit on the amount of energy which could have been added or removed from the CMBR. Figure 2.5 is a plot of the results of such a calculation. The vertical axis specifies the fractional energy, where ΔU is the limit on the energy created or modified by these processes, and U is the energy in the CMBR. The horizontal axis is redshift +1, or the inverse linear scale factor for the Universe (1 on this axis represents the current size of the Universe, 10^3 is when the Universe was 1000 times smaller). The line increases dramatically at redshifts above 10^6 because we have entered the era where the Universe is so hot and dense that the forces which drive the CMBR into thermal equilibrium are exceedingly strong. This has the effect of erasing any modifications to the Planckian form and therefore decreases our ability to limit these modifications. The FIRAS observation implies that less than 0.03% of the energy in the CMBR is out of thermal equilibrium starting from a time when the Universe was roughly 1 year old. In a sense, on the largest scales and over

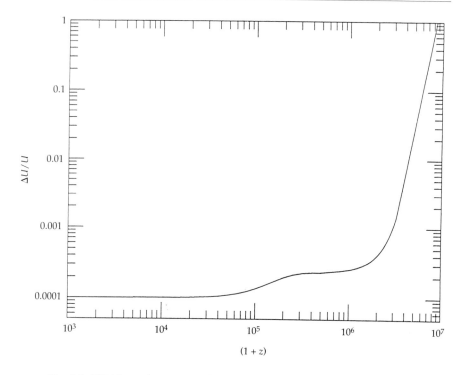

Fig. 2.5. FIRAS results converted to $\Delta U/U$. The implied fractional energy deviation from a blackbody curve at different redshifts.
(COBE Science Team/NASA)

much of its life, the Universe is as simple as the glass blower's furnace. What is also remarkable is that a spectrum that is predicted by theory based on laboratory measurements should so well describe the dominant component of the Universe.

Ever since the discovery of the CMBR, there has also been an intense effort to measure its brightness variation over the entire sky. Since this variation probes the distribution of matter in the early Universe, it has long been recognized that finding deviations from uniformity (anisotropies) would provide important clues to the evolution of the rich structures that we see in the sky today (finding perfect uniformity just means that one has not looked with sufficiently sensitive instruments). In spite of major efforts by many groups over the past two decades, only a 'dipole anisotropy' had been detected until recently. This effect shows that half of the sky is brighter than the other half by one part in 1000. The center of the bright part of the sky is roughly at right ascension 11 hours and declination $-6°$ (in the constellation

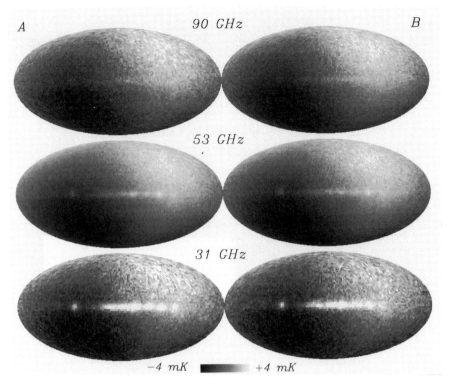

Fig. 2.6. DMR maps with dipole. One-year of DMR observations. There are two independent radiometers at each wavelength. The panels are labelled by the frequency : 90 GHz is 0.33 cm, 53 GHz is 0.57 cm, and 31 GHz is 0.94 cm. (COBE Science Team/NASA)

Leo). This dipole anisotropy is largely due to the Doppler shift from the motion of the Earth relative to the CMBR and implies that we are moving towards the constellation Leo at approximately 360 kilometers per second, about 0.1% of the speed of light.

Recently, the COBE team announced the discovery of brightness variations at a level of 1/100 000th of the brightness of the CMBR. Figure 2.6 gives a set of sky maps from which this conclusion was reached. These are maps of the brightness of the sky at wavelengths 0.94, 0.57 and 0.33 centimeters, with two sky maps per wavelength because there are redundant receivers in the COBE spacecraft. The dominant feature is the dipole anisotropy from the motion of the Earth relative to the CMBR. Figure 2.7 shows the same maps with the dipole anisotropy removed. The coordinates of the maps are such that the Galactic plane lies in the middle, as can be seen clearly in the 0.57 centimeter map. The CMBR variations are not clearly visible in these maps

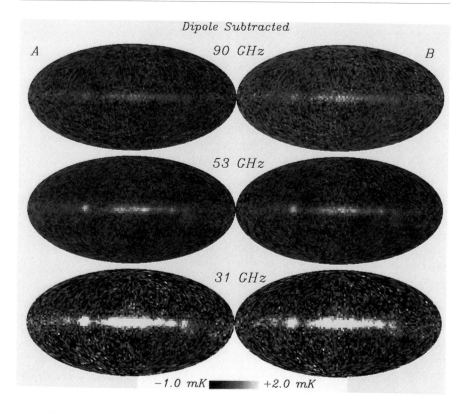

Dipole Subtracted

Fig. 2.7. DMR maps without dipole. One-year of DMR observations with the dipole anisotropy removed, showing residual Galactic radiation. The Galactic plane divides the top and bottom halves of each ellipse. The remaining bumps are mostly measurement error, but a few of the variations away from the Galactic plane are true CMBR variations, some showed on the two-year maps. (see also the cover.). These variations will stand out in greater detail when the full four-year maps are completed.
(COBE Science Team/NASA)

because they are about the same size as the variations arising from noise in the instrument. This unfortunate fact prevents us from pointing at a particular place in the map and marking an 'object' as a definite CMBR variation. We are currently analyzing additional data which will make these variations more visible to the unaided eye. Meanwhile, we can perform statistical analyses of the characteristics of these maps to detect and characterize the variations. Figure 2.8 shows the fundamental result in the form of an auto-correlation function, which is a statistical measure of the amount of variation in the map. In particular, it describes how much pairs of points which are

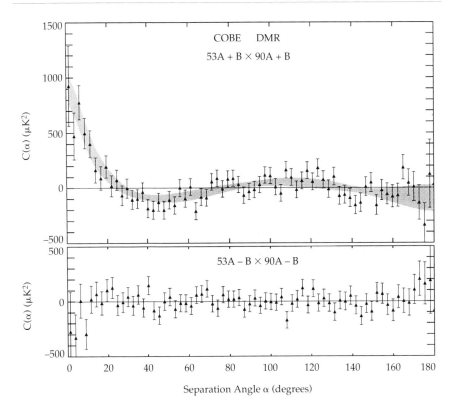

Fig. 2.8. DMR autocorrelation function. This is the fundamental result from the DMR experiment which characterizes the variations seen in the sky. The band shows the prediction of inflation theory.
(COBE Science Team/NASA)

separated by a specified angle (the x-axis in the plot) are similar to each other. If there were no variations in the sky brightness, we would expect to see a noisy line around 0. The solid line is a fit to a popular model of the auto-correlation function as prescribed by several theories.

Shortly after the COBE announcement, a ballooning group reported measurements on a smaller patch of the sky (Fig. 2.9) that match the brightness variation characteristics reported by the COBE group. This balloon measurement is at a wavelength which is three times smaller than that of the most sensitive DMR channel (0.57 centimeters), allowing for detailed checks of the source of the variations. As we will discuss below, there are many possible sources of variations at these minuscule levels, and the detection of similar signals by two different experiments at vastly different wavelengths

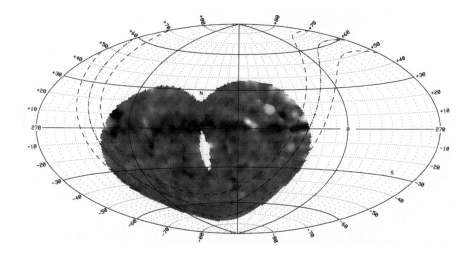

Fig. 2.9. Far-IR survey map. These data are taken at a factor of 3 shorter wave-length than the most sensitive DMR channel. They are also dominated by instrument noise, but a statistical comparison with the DMR maps confirms the DMR discovery (regions where the DMR maps tend to have higher intensity also have higher intensity in this map, and vice versa).
(Ed Cheng)

strengthens the argument that the effect is real, and not an artifact of the design of either experiment or of some other source of emission in the sky.

Both of these experiments are sensitive to variations which are large compared with the horizon size at the surface of last scattering. The DMR observed the sky with a 7° beam and the balloon experiment observed with a 3.8° beam (the disk of the Sun or Moon are both about 0.5° across). This implies that the observed variations are the imprints of processes occurring before the surface of last scattering, and not a result of physical processes occurring at that time.

In anticipation of the recent results, a number of groups have already begun programs to investigate the CMBR isotropy at angular scales finer than those studied by the COBE. Because these scales are smaller than the horizon size at the surface of last scattering, some theories predict a specific increase of variations with decreasing angular scale, starting at about 2°. These theories use the COBE results on the large scales (10° to 180°) and make predictions based on the physical processes that may be occurring at the surface of last scattering. Measurements at these smaller scales would provide an important confirmation of the COBE results in addition to checking the predictions of the theories we rely on to understand them.

HOW DO WE SEE IT?

One might ask why it takes two decades to come up with what may appear to be relatively simple observations of the CMBR. The answer is, of course, that these are very difficult measurements to make. Most of the necessary techniques had to be learned over these past two decades.

The peak brightness of the CMBR radiation occurs at a wavelength of about 1 millimeter, near the far-infrared region of the electromagnetic spectrum. Since a Planck spectrum has a relatively broad distribution of brightness near its peak wavelength, it is possible to study the CMBR down to the microwave regime (wavelengths of about 10 millimeters or so), but it is not possible to look too far away in wavelength. For instance, visible light is at a wavelength of 0.0005 millimeters. Even at wavelengths much longer than this, the CMBR has become so faint that other sources such as stars are dominant.

The first problem that one encounters in even crude studies of the CMBR is that the atmosphere is a severe impediment. When attempting to make measurements of the CMBR in the presence of atmospheric emission that is almost as bright, one must invent clever schemes to remove the signal arising from atmospheric emission, as was done for early studies from the ground. Still, the larger the competing effect, the harder it is to do a good job of removal. The problem is not just with the brightness of the atmosphere, but also in the variability of this brightness over the time needed to make the observations (typically many hours to days). This led to experiments on mountain tops and remote areas such as Northern Canada and Antarctica, where we can minimize the contribution from water vapor, the most variable and hardest to predict component of atmospheric emission. Such experiments are relatively easy to perform, and are typically completed in about two to five years. The next steps in complexity are to try to get away from the atmosphere by using balloons, rockets and satellites.

Balloons are the simplest of these high altitude platforms. At their limit, they can bring payloads of up to 3000 lbs (1400 kilograms) to an altitude of approximately 140 000 feet or 6600 meters (about 4 times the altitude of a commercial jet aircraft), where the atmospheric pressure is reduced by a factor of 200. Balloons can provide several days of observation time per launch during low wind seasons. More typically, a single successful flight produces one night (6–10 hours) of data. The equipment is subject to a low pressure

(0.005 atmosphere), low temperature (−50°C) environment, and must be suitably designed and tested. At the end of the flight, the payload is returned to Earth via parachute. Such experiments have typically taken from four to six years from concept to launch.

Some experiment designs prefer a vacuum environment which can only be provided at even higher altitudes. Sounding rockets can take small payloads of up to 500 lbs (230 kilograms) to an altitude of well over 330 000 feet (100 000 meters). At these altitudes, there is practically no atmosphere, but the trajectory of the rocket limits observation time to about 10 minutes. This brief time limits the raw sensitivity of the measurements, but may be a valid trade off in some cases. A related disadvantage for such brief measurements is the inability to perform extensive tests to ensure that the data represent true signals from the sky and are not contaminated by the instrument. The equipment must survive launch vibration stresses in addition to the space environment. Because of this added complexity, the typical development time is a little longer than that of balloon payloads.

The satellite platform can virtually eliminate atmospheric concerns, as well as provide for a long observation time. However, because of the launch environment and the long life involved, the spacecraft tends to be more complex and the development time is consequently longer. An example is the COBE, which took 15 years to evolve from concept to launch. It is worthwhile to note, however, that this extra effort also brings unique capabilities. An important aspect of the longer observation time is the ability to run tests which check the validity of the data. Such tests may include measuring the sensitivity of the apparatus to a variety of potentially bothersome effects ranging from atmospheric emission to the Earth's magnetic field. When the observation time is short, many of these tests are not possible, and the experiment runs the risk of being flawed. Another advantage of a space mission is that the entire sky can be covered using an appropriate orbit. Combined with instruments which provide measurements at many different wavelengths, such missions can yield full-sky, multispectral data sets which are crucial for establishing the nature of the detected signals.

A new approach of long-duration ballooning from Antarctica holds the promise of satellite-like stability and substantial observing time while allowing for (relative) simplicity of execution. One of the problems of ballooning over many days is that the helium gas which provides the lift tends to escape during, the thermal variations experienced during sunrise and sunset. This makes it difficult to keep a balloon stable at the required altitudes for more than a few days at a time. During the Antarctic summers and winters, the Sun is either always up or down, respectively, eliminating this concern and allowing for flights which can last at least two weeks. Logistical constraints give a strong preference to the Antarctic summer, which is slightly inconvenient because the Sun is a possible source of stray radiation. However, its position is well-known and protective shields can be installed. This approach offers an extended period of stable operation in an environment which is practically free from atmospheric emission. Many more checks of instrument error sources can be completed in a two week period (compared with the few hours available from an overnight flight). The shorter observation time compared with satellites can be partially overcome by using more advanced detectors. This phenomenon is a little strange, but is a common consequence of the long development time of the space missions. At the time of design, which may be a decade before launch, the most mature technologies are usually chosen in order to minimize the cost and risk of development. By the time the mission is launched, the technology used may be somewhat older than that which is possible for programs with a shorter development time.

A central technological question in our measurement abilities is the concept of sensitivity. It is much easier, from a sensitivity standpoint, to detect the CMBR radiation than to detect variations which are 1/100 000'th of the brightness of that radiation. Depending on the goal of an experiment, we can optimize by looking at available technologies, and by increasing the observation time (noise limited sensitivity improves as the square root of the time spent on a particular observation). The earlier experiments almost exclusively used microwave technology, which is a relatively well-developed technology for communication and radar. During the past decade, improvements in far-infrared detectors have led to measurements near the peak brightness of the CMBR, where one can study more closely the detailed shape of the spectrum, as well as its variations. As new technologies develop, CMBR experiments will no doubt evolve to use and also help refine them.

After having solved the atmospheric emission and sensitivity issues, an experimenter will take data and discover yet another impediment to measuring the CMBR, that of astronomical foreground sources. I made the claim that, after the era of decoupling, the photons traveled to us unimpeded. While this is approximately true, the fact that there are galaxies and clusters of galaxies between us and the origin of the CMBR light allows for many sources of interference. These sources may not modify the CMBR itself, but may cause additional features in the sky brightness maps which obscure or add to the CMBR and therefore must be understood and removed. Our ultimate understanding of the CMBR is limited by our ability to understand these sources.

Most of these problems have been experienced during the past two decades, and solutions are incorporated into current CMBR experiments. For problems like atmospheric or Galactic emission, we know (at least crudely) how this emission varies with wavelength. Such contributions can be approximately removed using models based on measurements at multiple wavelengths, and can be minimized by observing at wavelengths where the CMBR signal is relatively more prominent. Nevertheless, designing and optimizing these measurements is an ongoing challenge as the level of sensitivity increases, and as new problems and contaminating effects are uncovered.

HOW DO WE MAKE THESE OBSERVATIONS CREDIBLE, AND HOW DO THEY RELATE TO OTHER OBSERVATIONS?

The farthest of objects, which are the ripples of matter revealed by variations in the CMBR, must be related to the structures which we see in the sky today (galaxies, clusters, voids, etc.). One of the major challenges in cosmology is to relate the sizes of these structure 'seeds' to quantitative measures of the current distribution of matter. These seeds are no more than extremely small variations in the distribution of matter at the era of decoupling, and provide a starting point for all theories of large-scale structure evolution. The currently observed distribution of matter (as well as we have understood it) provides the endpoint. The game is to provide a plausible scenario that can tie these two together.

We have data on the current distribution of galaxies within a sphere of roughly 3×10^8 (300 million) light-years radius. Assuming that the Universe has been around for about 15×10^9 years (15 billion), this is the most recent 2% of the lifetime of the Universe. In the human adult analogy, this corresponds to the past 6 to 8 months of one's life. As I mentioned before, the era of decoupling occurred at a time which corresponds to only a few hours after conception. This leaves a large gap of time for which we have little observational data, and which any plausible theory of evolution for the Universe must fill in.

This effort has not been entirely successful in providing details, and is currently the focus of intense investigation. If we believe the general framework we have been building, we should be able to create the structures observed today from the ripples observed at the epoch of decoupling in the time that has elapsed since the epoch of decoupling. It turns out that the primeval ripples may be too small (in amount) to have developed into the required structures when using the simplest assumptions for the nature of gravity and the amount and constitution of matter in the Universe.

A popular approach is to invent a new class of 'dark matter,' which is not visible (or at least, not yet detected) and which floats around the Universe 'seeding' the development of the large-scale structures seen today. There can be dark matter made from normal matter, but this matter is found to move around too fast to encourage enough clumping in the given time. If we add more to get more clumping, then there is an uncomfortably large amount of mass in the Universe. This material is called 'hot' dark matter because it moves so quickly. Naturally, such a problem leads one to hypothesize another kind of 'cold' dark matter which can provide the necessary clumping, but must have special properties so that it does not move around. Such matter is likely to be a new type of particle that has not yet been detected, or a result of some property of existing particles that we do not fully understand. While such assumptions may seem contrived, these theories have had reasonable success at predicting structures which closely resemble the observations, and have motivated a few ongoing attempts at detecting dark matter in the laboratory.

A different approach to creating large-scale structures is through the nature of space itself. In some high-energy particle physics theories which are relevant to the conditions in the hot and dense phase of the early Universe, it is possible to create 'defects' in space while it is expanding and cooling. These defects have the effect of being immense sources of gravity in line-like (cosmic strings) or more complicated structures (cosmic 'textures'). These exotic structures would then provide the attractive force needed to develop large-scale structure in the required time. Such theories have also enjoyed some success at matching the observations.

Alternate scenarios have also been proposed that question the validity of the physics we apply to the early Universe. For example, suppose that gravity does not behave exactly as we understand it when applied to objects comparable to the size of the Universe. We would then be able to construct a new model of gravitation on these large scales which would match the observations. One such modification of gravitational theory had originally been proposed by Einstein as a simple way to create a static Universe, but was abandoned after the Universe was shown to be expanding. This same type of modification is becoming interesting again.

Thus, as you can see, it is not yet entirely clear how the pieces of the puzzle fit together in detail. In order to distinguish amongst these competing models, we need to get a clearer picture of how the amplitude (amount) of the apples changes with scale size. The COBE observations are a major breakthrough, offering a very clear picture on the largest of scales, but does not constrain the theories for scale sizes smaller than the largest structures observed today (the smallest bump that COBE can observe at the surface of last scattering has evolved to a size of about the largest observed structures today). As more of the intermediate and small-scale CMBR anistropy measurements return data, we will get a better idea of which of these scenarios works best.

CAN WE LOOK FARTHER THAN THE SURFACE OF LAST SCATTERING?

The surface of last scattering interpretation, if correct, places a severe constraint on our ability to see structures that are farther away (or that occurred when the Universe was younger). However, this constraint applies only to light or other electromagnetic radiation. Using another information carrier, such as neutrinos or other elementary particles yet to be discovered, may allow for probes into the more distant past. The thinking involved would be similar to that developed here. That is, at some time in the past, the Universe must have cooled sufficiently to allow these particles to condense out of thermal energy, and in a sense become 'real.' At an even lower temperature, the interaction of these particles with the Universe could have been decoupled, and thus provide a picture of the distribution of matter at that time. However, this idea is clearly very speculative since we have only begun to explore the CMBR and the surface of last scattering, and we have not yet even conceived a method to probe these other components of the background radiation.

WHAT HAPPENED BEFORE THE BIG BANG AND WHAT IS AT DISTANCES FAR GREATER THAN THE HUBBLE DISTANCE?

The world view that we have been discussing requires that there is a distance that we cannot see beyond, and specifies an age of the Universe. We might be tempted to think that, by being sufficiently clever, it may be possible to deduce what happens at even farther distances based on what we are allowed to see. A model of the unobservably far parts of the Universe can be tested against what is seen at the farthest of places. We require, of course, only a model of how things develop as the distance changes. This is analogous to seeing an airplane coming over the horizon and asking where it came from. Given measurements of its speed and direction (say, by radar), we can deduce likely answers, but the precision will depend on how far beyond the horizon we want to study. Unfortunately, it is very unlikely in the currently accepted scenario for us to deduce what happened before the Big Bang since we do not know how to model the behavior of physical objects (unlike the airplane, whose behavior we can predict using mechanics). It seems that most conjectures lead to the conclusion that any physical information from before this time, if it is even meaningful to ask this question, is erased by having the Universe go through a hot, compressed state. We must wait for progress in the field of quantum gravity, the study of gravitational phenomena in small, dense regions, in order to understand better the physical laws which apply under these conditions.

A similar question often arises about what is 'outside' our Universe, or just what exactly are we expanding into? The short answer is that there is no outside and that it is the space in the Universe itself which is expanding. It is possible to postulate other Universes which exist outside of ours or are started by a separate 'Big Bang,' but since we do not know how to think about interactions with these other Universes, it is hard to pursue this line of reasoning. These are simply areas in which currently understood physics has internally consistent, but perhaps not very satisfying answers.

When confronted with these more 'fundamental' questions, we must be a little pragmatic in our approach, and perhaps also humbled by our lack of understanding. There really are no satisfactory answers to these kinds of questions until our understanding of our own Universe, as defined by the volume within roughly a Hubble distance, is more complete. One can speculate endlessly about the possibilities, but unless we can make definite predictions and perform physical tests, such thoughts must remain as curious speculations.

EPILOGUE

In the course of studying the farthest of objects and the nature of the early Universe, there is a sense of *déjà vu*, since the commonly accepted scenario is not unlike some beliefs which were described, millennia ago, in various creation stories. One is frequently led to the question why this similarity exists and whether it is due to coincidence, a fundamental part of how the human mind works, or if some other phenomena outside the realm of current scientific knowledge are at work. This is most likely a subject we will ponder for many generations to come. However, the science of cosmology, which is based on testable, quantitative predictions, is now at the intriguing point where we can start to discuss, in some detail, what appear to be the physical processes following the birth of our Universe. This knowledge, while still limited, will undoubtedly form one of the pillars for our understanding of the context in which human beings exist in the Universe.

CHAPTER THREE

•

QUASARS

•

PATRICK S. OSMER

Quasars, which can be a thousand times brighter than an ordinary galaxy, are the most distant objects observable in the Universe. How quasars produce the luminosity of 10^{13} suns in a volume the size of the solar system continues to be a major question in astronomy. Distant quasars are very rare objects whose study has been blocked by their scarcity. Recent technical advances, however, have opened new paths for their discovery. Forty quasars with redshifts greater than 4 have been found since 1986. Redshift 4 corresponds to a light travel time of more than 10 billion years. As a result, we are now able to probe the epoch shortly after the Big Bang when quasars may have first formed and to study the universe when it was less than a tenth its present age.

Quasars were one of the main discoveries thirty years ago that revolutionized astronomy. While they and the black holes thought to occur in their centers have become household words today, quasars are as enigmatic in many ways as they were when first discovered. Whatever their nature, they offer us views of the Universe never before seen, especially at distances far beyond what astronomers of the previous generation expected to see. In this chapter I wish to review briefly their history, how extraordinary their properties are, and how they serve as probes of the Universe to nearly as far as the visible horizon. Then I describe the case for how the population of quasars has evolved with cosmic time and how recent technical advances have led to the discovery of very distant quasars, the studies of which can tell us whether or not we have seen back to the epoch of quasar formation.

DISCOVERING QUASARS

Quasars burst upon the astronomical scene in 1963 when Maarten Schmidt succeeded in identifying the spectral lines in the newly identified stellar object at the position of the radio source 3C 273. (Its name indicates that it is the 273rd object in the 3rd Cambridge catalogue of radio sources.) Schmidt showed that the lines came from normal atoms but were shifted to longer wavelengths by 15.8% of their regular values, something entirely unprecedented for a starlike object of 13th magnitude. If the shift arose from the expansion of the Universe, 3C 273 was intrinsically much brighter than an entire galaxy and was at a distance of some 3 billion light-years, properties far beyond the expectations of research astronomers at that time. For example, a star like the Sun would only be at a distance of 1400 light-years to have an apparent magnitude of 13; a very luminous galaxy containing 100 billion suns would be at 400 million light-years.

Schmidt's result sparked feverish activity in the field. It is fascinating and sobering to look back at the proceedings of the First Texas Symposium on Relativistic Astrophysics held at the end of 1963: fascinating to see how many of the basic properties and problems of quasars had been established in less than a year, sobering to realize how difficult it has been to make substantial progress in verifying some of the main phenomena that occur in quasars. Nonetheless, our knowledge about quasars and what they tell us about conditions in the far reaches of the Universe when it was less than a tenth of its present age has expanded dramatically since 1963. Most astronomers agree that quasars do serve as beacons that illuminate lines of sight stretching out more than 20 billion light-years at the present scale of the Universe. With spectroscopic techniques we can deduce much about the distribution of intervening objects such as gas clouds and galaxies that are otherwise too faint to be seen.

Not all astronomers agree that quasars are at the distance indicated by their redshifts, in which case the redshifts would be the result of physical processes not yet understood. The debate on this topic has been covered in a previous session of the American Association for the Advancement of Science and a resulting volume by George Field, Halton Arp and John Bahcall (see Reference list). For the remainder of this chapter, I shall assume that quasars are at the distances corresponding to their redshifts.

QUASARS AS EXTRAORDINARY OBJECTS

The observed properties of quasars on which there is general agreement are:

(1) they have starlike nuclei in optical images;

(2) those quasars that are detectable radio sources show structure ranging from milli arcsec scales to tens of arcsec, often with evidence for apparent expansion of the most compact structures on time scales of years;

(3) radiation of similar power in energy bands ranging from gamma rays through the optical and infrared to the radio region of the electromagnetic spectrum;

(4) redshifts of spectral features from 0.1 of the rest wavelengths to the currently largest value of 4.9;

(5) variability of apparent brightness in different bands on time scales as short as days to weeks;

(6) luminosities (if at the distances indicated by their redshifts and assuming isotropic radiation) up to 10^{14} suns; and

(7) stored energy of up to 10^{60} ergs in high energy electrons in the associated radio sources.

The large energy output from quasars combined with the small size implied by the short time scale of variability (light days, not that much bigger than the solar system) made it immediately clear in 1963 that quasars were not like ordinary stars or galaxies, nor were nuclear fusion reactions likely to be adequate sources of energy. Perhaps the area of widest agreement today is that gravitation is the main power supply. Martin Rees reminds us, 'any gravitationally-powered source which releases more than 1% of its rest mass energy contracts unstoppably; and a collapsed object, once formed, offers a more powerful and efficient power source than any precursor system.' Thus, the massive black hole concept for quasars has arisen, with the schematic idea being that a quasar consists of a 10^8 solar mass black hole located at the center of a galaxy. Even though no light can escape from the hole itself, matter falling towards the hole becomes very hot and radiates a tenth or more of its rest energy *en route*. The vicinity of the black hole is not a hospitable place. We know from observations that gas motions of 10 000 km/s or more occur even at distances greater than a light-year from the hole. Closer in, the temperature exceeds a billion degrees; within a light-day of the hole, effects of general relativity become important. Rees' diagram

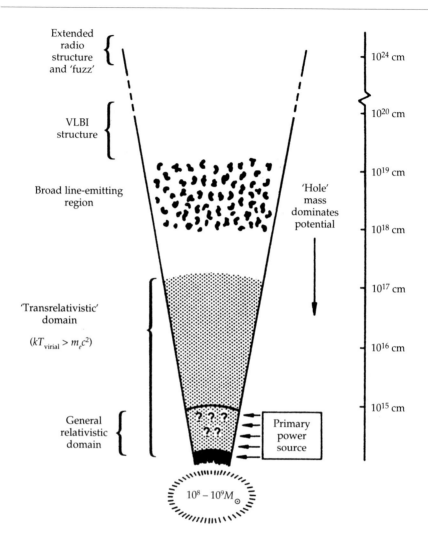

Fig. 3.1. Schematic illustration of a quasar, produced by M. Rees.

(Fig. 3.1) shows that the quasar phenomenon extends over a range of length-scales of 10^{10}. The radius of the hole itself is 16 light-minutes or twice the distance from the Earth to the Sun, while the largest radio structures associated with quasars extend to millions of light-years.

Given the extreme range of conditions that exist in quasars and that they in general do not follow equilibrium processes, it is understandable that they have been difficult to model. Clearly our understanding of stellar astrophysics has been greatly aided by the fact that stars are in equilibrium and that the ideal gas law applies for much of their lifetime. Quasars continue to present a challenge to theorists in virtually all aspects of their structure and properties.

QUASARS AS PROBES OF THE UNIVERSE

Even if our understanding of quasars is incomplete, their great luminosity gives us our best and in some cases only view of the Universe at large distances and early epochs. Table 3.1 illustrates a few examples of the scales of the Universe that we wish to discuss. It is constructed for the Einstein–de Sitter cosmological model, in which the amount of matter in the Universe produces a deceleration of the expansion just sufficient to stop it at infinite time. This is the type of model that results from the so-called inflationary cosmology pioneered by Alan Guth. For the parameters used, the age of the Universe is 13 billion years. Thus, the most distant quasar known is being seen as it was 12 billion years ago, only a billion years after the Big Bang. Note that the quasar is now at a distance of 23 billion light years because of the expansion of the Universe in the time after the light now being received was emitted. By comparison, the familiar galaxy M31 in Andromeda and even the great cluster of galaxies in Virgo are very close neighbors.

Although we are now accustomed to observing quasars at great distances from Earth, we should not forget that the simple ability to do so means the Universe is extremely transparent to the passage of light when we look away from the plane of our own Milky Way Galaxy. Within our galaxy, we cannot even see 25 000 light years to its center because of the large amount of interstellar dust that occurs in the galactic plane. In principle, the occurrence of a single such galaxy along the line of sight to a quasar could block its visible

Table 3.1.

(For a Friedmann model with $H_0 = 50$ km/s/Mpc, $q_0 = 0.5$)

Object	Distance (Light-years at present time)	Redshift (from expansion of Universe)	Light travel time (for radiation now reaching us, in years)
M31	2×10^6	1×10^{-4}	2×10^6
Virgo	65×10^6	3.3×10^{-6}	65×10^6
Nearby quasar	1.8×10^9	0.1	1.75×10^9
Intermediate distance quasar	11×10^9	1.0	8.4×10^9
Most distant known quasar	23×10^9	4.9	12×10^9

light completely from our view. Even neutral hydrogen in intergalactic space could effectively obscure light at wavelengths shortward of the main Lyman alpha emission line in quasars as shown by J. Gunn and B. Peterson in 1965. That such effects do not occur is an indication that the Universe does not contain a large amount of absorbing material, an important topic to which we will return below.

Although transparent, the Universe at large redshift is by no means empty. The spectra of distant quasars contain many absorption lines that arise from diffuse clouds of gas and what appear to be different kinds of galaxies lying along the line of sight. These objects are in general too dim to be seen by their own emission, but they produce characteristic signatures in the spectra of background quasars. Fig. 3.2 illustrates the spectrum of a distant quasar showing the different types of absorption lines. The first systems to be identified contained lines of familiar elements like carbon and magnesium and are presumed to arise in the gaseous halos of galaxies. After a number of quasars with redshifts greater than 2 were discovered, in which the strongest transition of neutral hydrogen is shifted into the visible part of the spectrum, it was realized that all such quasars showed a large number of weak absorption lines at wavelengths shortward of the transition, which is Lyman alpha at 121.6 nm. R. Lynds argued in 1971 that these lines were most plausibly Lyman alpha in multiple clouds of low column density along the line of sight; they are now familiarly known as the Lyman alpha forest. In about 20% of distant quasars extremely strong absorption from Lyman alpha is seen. These clouds, with column densities in excess of 10^{20} atoms cm^{-2}, are thought to occur when the line of sight traverses the disk of a galaxy itself. Because the physics of absorption line formation is often more tractable than for emission lines, we in some cases can deduce more about the physical conditions in the absorbing clouds than for the emission regions of the background quasar itself. For example, the absorption lines are used to study the chemical abundances of interstellar gas at early epochs. It already appears as if dust is not present in the distant absorbing systems in the same ratio as for clouds in our own galaxy. Otherwise, we would not be able even to see the quasar for the obscuration that should be produced in the systems with high column density of Lyman alpha.

THE EVOLUTION OF QUASARS

In a particular volume of space, there will be more faint quasars than intrinsically bright ones. The luminosity function is a quantitative measure of the

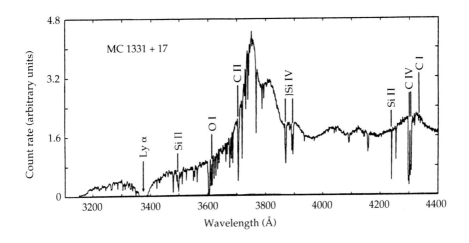

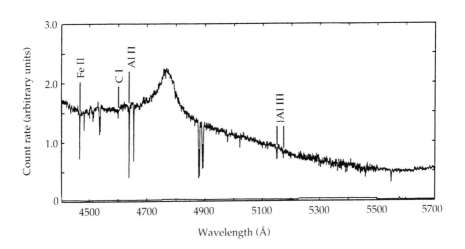

Fig. 3.2. Spectrum of the quasar MC 1331+17 obtained by F. Chaffee, J. Black, and C. Foltz. It shows the continuum from 3200Å to 5700Å, the emission lines (broad features rising above the continuum) typically seen in quasars, and a variety of absorption lines (narrow features going below the continuum). The broad absorption feature just below 3400Å is Lyman alpha from a high density absorbing cloud. There are many weak, unmarked, Lyman alpha absorption features shortward of 3700, the so-called Lyman alpha forest lines. Finally, there are a variety of strong, narrow absorption features from elements such as carbon, oxygen, silicon, and aluminum that are marked in the spectrum. They arise in intermediate density clouds along the line of sight.

density of quasars according to their brightness. More distant quasars appear dimmer than nearby ones of the same luminosity. When a complete sample of quasars is observed in some part of the sky to a particular limit in apparent brightness, the resulting distribution of objects by redshift and brightness will depend on the luminosity function, how it may change with redshift, and the cosmological model adopted (although the observations are usually analyzed in a form to reduce the sensitivity of the results to the adopted cosmological parameters). Considerable effort has been made to invert the problem so that the underlying properties can be derived from the data. Such techniques have a distinguished history in astronomy and were used, for example, to deduce the structure of our own galaxy.

In 1968, Schmidt realized that even the small sample of radio quasars available to him contained surprisingly strong evidence that quasars occurred much more frequently at higher redshifts and therefore earlier times. There were far too many quasars with redshifts near 2 in his sample compared to what would be expected if their space density were uniform over the interval from 0 to 2, that is, from the present epoch back to when the Universe was 20% of its present age.

Schmidt's work opened up what has become a major research topic in quasars. Today there is general agreement that the data can be interpreted as showing an increase in the space density of quasars by a factor of as much as a thousand in going back to redshift 2. However, two major questions being actively debated are: (1) Is the density higher at redshift 2 or were quasars brighter then? and (2) Does the density decline significantly at redshifts higher than 2?

Soon after Schmidt's first work on the density evolution of quasars, alternative interpretations began to be considered, especially the possibility that the space density of quasars could be constant but their luminosity increased significantly at higher redshifts, which is the luminosity evolution model. The debate on density v. luminosity evolution continues to this day. The question is extremely important because of the implications the results have for the nature and lifetimes of quasars. In the density-evolution picture, quasars would be frequent, but shortlived phenomena in which most bright galaxies would have participated at earlier epochs. In the luminosity evolution picture, quasars would be a rare event occurring in perhaps 2% of all galaxies, but lasting over most of the lifetime of the galaxy. Thus, the statistics of faint quasars carry potentially crucial clues as to their physical nature. At the same time, we should be aware of the limitations of the data and not necessarily expect that counting quasars alone will settle the matter. In any event, this topic is worthy of more extensive treatment than can be given here, and the reader is invited to pursue it separately.

The second question, does the density of quasars decline at redshifts greater than 2, may appear more straightforward to answer but in practice has not been easy, owing to the small numbers of quasars at the highest redshifts and the difficulty of efficiently separating them from stars. The question of a possible cut off in the redshift of quasars was raised by Schmidt and A. Sandage in the early 1970s and its reality has also been debated ever since. However, recent technical advances have led to substantial progress in the field.

Because quasars are so intrinsically luminous, they will be well within the detection limit of groundbased telescopes even at high redshifts. For example, 3C 273, the first quasar discovered, would have a brightness of approximately 20th magnitude if it were at redshift 5; modern telescopes can detect sources nearly a thousand times fainter and can obtain spectra of objects ten times fainter. The problem lies in deciding which objects are good candidates to be distant quasars.

FINDING DISTANT QUASARS

A wide field instrument such as the UK Schmidt telescope in Australia can photograph 36 sq deg of sky on a single plate, which will reach to 22nd magnitude and contain more than 200 000 images of stars, galaxies, and quasars. Since a quasar is by definition starlike, a single photograph alone is incapable of indicating which images are stars and which are potential quasars. In anticipation of the results described below, we now know that such a photograph will contain of order 1000 quasars in all, but that only a handful will have redshifts greater than 4. To find them is a true needle-in-a-haystack problem that requires great care in image and signal processing.

Quasars are operationally distinguished from stars and galaxies by some combination of their spectral energy distribution, emission at radio and X-ray wavelengths, point source appearance on photographs, lack of detectable motion on plates taken at different times, and their variability in light output. All of these characteristics have been used to identify different types of quasars. For the search for quasars at the highest redshifts, it has turned out that they are best distinguished by their spectral energy distribution and point source appearance.

Two techniques are currently being used successfully: multicolor imaging and very low resolution spectroscopy. Both need to cover large areas of sky (tens of square degrees) and reach to faint magnitudes. Also required are large format electronic detectors, or, in the case of photographic plates, high speed plate scanning machines. Both techniques require sophisticated image processing techniques, especially if they are to yield quantitative results.

The multicolor technique is based on direct images of the sky taken through filters of different color (spectral passband). In the case of photographic plates taken with the Schmidt telescope mentioned above, it is first necessary to make a digital scan of each plate, mapping the brightness of each picture element on the plate. Photographic plates are still essential in astronomy because their large area gives them an enormous number of picture elements, 10^9, which more than compensates for their low detective efficiency compared to electronic devices such as charge coupled devices (CCDs). A key step in being able to utilize plates for large area searches has been the development of rapid plate scanners, such as by E. Kibblewhite, which are able to process a plate in less than a day. Equally important, however, has been the perfection of efficient software to identify objects in the scans automatically, separate stars from extended objects such as galaxies, and derive brightnesses for all selected objects in the different colors. Finally, S. Warren, working with P. Hewett and M. Irwin and others, has developed methods to find stellar, that is, not extended spatially, objects that do not have the colors of ordinary stars and have a high probability of being quasars. To do this last step, Warren was able to use multiple exposures of the same field with the same filter to reject the inevitable flaws and defects on the plates that would otherwise swamp a list of candidate quasars with spurious objects. In addition, he incorporated the use of upper limits for the detection of objects that did not appear in all plates and therefore did not need to demand that a candidate be seen in all plates. This enabled the search for quasars to be pushed to even fainter limits; indeed the majority of the high-redshift quasars found were missing on at least one plate.

The multicolor technique works because it covers a considerable baseline in color, with five filters from 350 nm to 800 nm, and because normal stars have spectra quite different from quasars. Stars are thermal radiators whose spectra differ from black bodies in well understood and parameterized ways, the main differences arising from absorption lines and absorption edges in the continuum. Quasars do not radiate as normal thermal bodies, rather their spectra consist of a power law continuum with contributions from emission lines, possibly some thermal components at ultraviolet wavelengths, for example, and occasional incidences of strong, broad absorption lines from gas flowing out of the quasar at velocities up to 0.1 the speed of light. By examining the distribution of the photographic data in multidimensional color space, Warren showed that stars fell in a well defined locus occupying perhaps 10% of the volume. He then selected quasar candidates as the most promising objects lying away from the locus of stars. Through followup spectroscopy at moderate resolution with large telescopes, Warren, Hewett, and P. Osmer (WHO) were then able to confirm which candidates were distant quasars and which were not. With this information Warren could perfect his search technique to the point that 25% of the best candidates now turn out to be quasars. These efforts produced the discovery in 1986 of the first quasar with redshift larger than 4.

Subsequently, WHO have assembled a sample of 100 quasars with redshifts greater than 2.2, a sample that includes three with redshift greater than 4. Figures 3.3 and 3.4 illustrate how the redshift 4 quasars are separated from

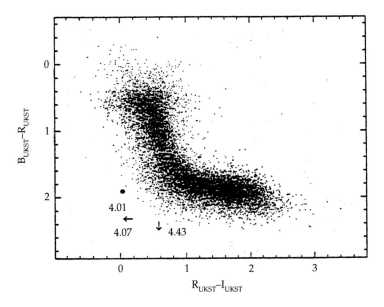

Fig. 3.3. Two color diagram by Warren, Hewett, Osmer, and Irwin showing how the three quasars with redshifts 4.01 (plotted as a blob to make it stand out), 4.07, and 4.43 (the arrows indicating that only upper limits have been measured) stand out from the main body of stars (the boomerang shaped cloud of points). The B (blue) – R (red) and R – I (photographic infrared) indices both increase as the color temperature of radiation decreases. The quasars stand out from the main body of stars because the quasars have excess light in the R band (relative to the I band) from the redshifted Lyman alpha emission and a deficiency of light in the B band from Lyman alpha absorption. *UKST* is the United Kingdom Schmidt Telescope in Australia.

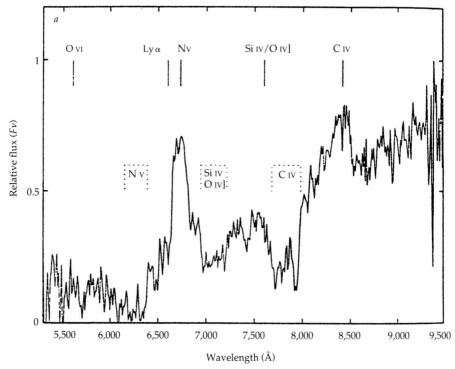

Fig. 3.4. Spectrum of the redshift 4.43 quasar, also by Warren *et al.* It shows redshifted emission lines of oxygen, silicon, and carbon in addition to the Lyman alpha line of hydrogen. There is also broad absorption from gas containing carbon.

stars in two color space and the image and spectrum of their object of highest redshift, 4.43.

A different, successful technique for finding high redshift quasars, developed by M. Smith and Osmer in the 1970s, employs a dispersive element in the optical system of the telescope to produce low resolution spectra for each stellar image. In this case quasars are directly recognizable via emission lines in their spectra or through the unusual shape of their continuum compared to stars. Conspicuous quasars are readily seen in visual searches of the images, especially when photographic plates are used. However, to reach fainter objects and to make a quantitative analysis of the results, electronic CCD detectors and image processing software are now being used. Schmidt, D. Schneider, and J. Gunn (SSG) are carrying out a large survey for high redshift quasars with the 5-m telescope at Palomar and CCD detectors. They have found a total of ten quasars with redshifts greater than 4, including one at 4.9, the farthest object in the Universe discovered to date.

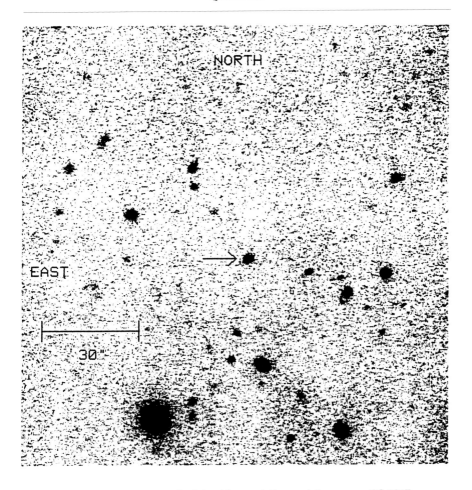

Fig. 3.5. Image of Schmidt, Schneider, and Gunn of the quasar PC 1247 +
3406, the most distant object known in the Universe.

Figures 3.5 and 3.6 show their image and spectrum of the quasar with
redshift 4.9. The spectrum has three distinctive features:

(1) the large redshift of the emission lines. Lyman alpha has been
shifted from its rest wavelength of 121.6 nm to 717 nm, with the other
lines being shifted proportionally. In terms of observations, this is a
shift from the shortest wavelengths that can be observed by the Hubble
Space Telescope to near the long wavelength end of the response
of optical detectors on groundbased telescopes. Indeed, it will be
difficult to discover quasars with redshifts greater than 6.5 with optical
detectors.

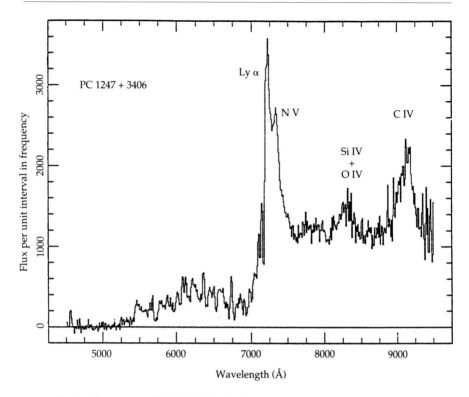

Fig. 3.6. Spectrum of PC 1247+3406 by Schmidt *et al.* Note that the emission lines are the same ones as seen in Fig. 3.4 but are narrower. Note also the steep drop of the continuum shortward of the Lyman alpha emission line, caused by absorbing clouds of hydrogen at large redshift.

(2) the apparent normality of the emission line strengths. The emission lines look remarkably similar in appearance to their counterparts in lower redshift quasars. Qualitatively, at least, this implies that the emitting gas in the most distant quasar has roughly the same chemical composition as ones closer to us. More importantly, this would mean that the gas around the quasar seen within a billion years of the Big Bang has been enriched to a similar level to that of the present time. This has important consequences for understanding how the elements were built up in the early stages of galaxy formation.

(3) a prominent drop in intensity at wavelengths shorter than the Lyman alpha emission line. This is caused by absorption from the increasing frequency of clouds of neutral hydrogen at high redshift and gives us information about the condition of intergalactic space at early Epochs in the evolution of the Universe.

THE SPACE DENSITY OF DISTANT QUASARS

The surveys of WHO and SSG now enable a new derivation of the space density of the most distant quasars, because they contain large enough numbers of objects and were carried out in a quantitative way. The latter point is essential for evaluating how the selection efficiency of the surveys depends on the redshift and brightness of the quasars being sought.

The WHO sample of 100 quasars is for an effective area of 43 square degrees on the sky. They estimate the probability of finding each quasar and employ it in a statistical methodology that yields a best fit parameterization of the data. They find (Fig. 3.7) that the space density of luminous quasars continues to increase beyond redshift 2 to a maximum near redshift 3.2. Then, at still higher redshifts, the density decreases by a factor of 5 by red-

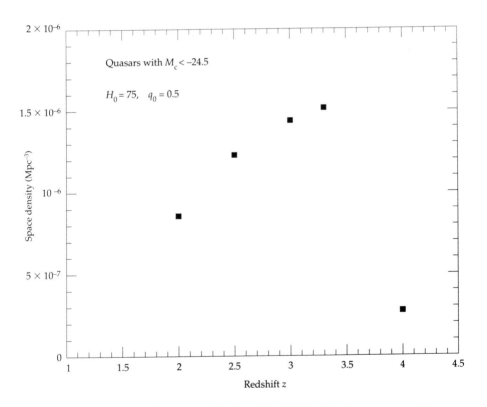

Fig. 3.7. The space density as a function of redshift for luminous quasars, as derived by Warren, Hewett, and Osmer. Their data indicate a continuing increase in space density from redshift 2 to 3.3 and then a falloff by a factor of 5 at redshift 4.

shift 4. Their result indicates that the peak of quasar activity occurred when the Universe was 12% of its present age.

The SSG survey involved the examination of several hundred thousand objects over 62 square degrees of sky. From their followup spectroscopy, they have compiled a sample of 141 quasars with redshifts between 2 and 4.7 for the analysis of the space density. Their work is completely independent of WHO, and their preliminary results are shown in Fig. 3.8.

Qualitatively the results look similar: a continuation of the space density from redshifts 2 to 3 and a decline by a factor of 3.2 per unit redshift at larger values. The agreement of the two well defined and independent efforts is the best available evidence in favor of the apparent space density reaching a peak near redshift 3 and declining toward higher values.

The result appears even more striking when plotted in the form of space density versus look-back time (Fig. 3.9). This indicates a strong peak at 0.8–0.9 the age of the Universe, or when it was 10–20% of its present age.

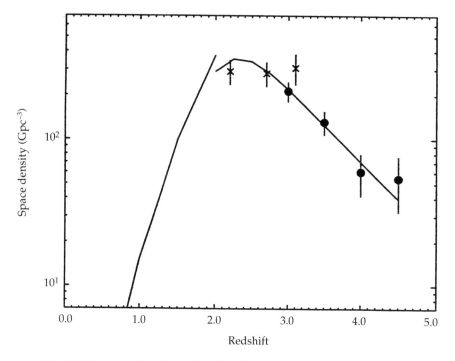

Fig. 3.8. The space density as function of redshift as derived by Schmidt, Schneider and Gunn. Their data points reach a maximum near redshift 3 and decline at higher redshifts by a factor of 3.2 unit redshift. The line for redshift < 2 is from Boyle.

Taken at face value, this would imply a sudden and sharp period of formation of quasars at an epoch not otherwise expected to be singled out. This epoch is clearly important to theories about the origin of quasars; at the same time it bears on the formation processes of the galaxies in which they occur.

For example, there is much evidence at current epochs that the close inter-

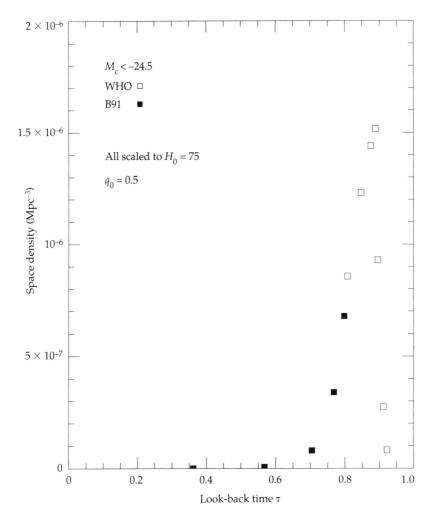

Fig. 3.9. The quasar space density plotted on a linear scale as a function of lookback time in units of the age of the Universe. Here the data of Warren, Hewett, and Osmer (WHO) at high redshift have been combined with the results of Boyle (B91) at lower redshift. Note the apparently narrow window in look-back time at 0.8 to 0.9 the age of the Universe when quasars were much more frequent than they are at present.

action or collision of two galaxies can lead to the formation of a massive burst of new stars. Subsequently the two galaxies can merge to form a single galaxy. It appears that the material in the galaxies can lose a large amount of angular momentum during the interaction and fall to the center of the merged object, where, of course, it is supposed that a black hole exists or is formed. This gives us one mechanism for understanding how material can fuel the quasar phenomena; it also illustrates the relation between galaxies and quasars.

In the meantime, M. Irwin, R. McMahon, and C. Hazard have devised a simple variant of the multicolor technique to find 23 new quasars with redshifts of 4 or larger, the largest sample yet compiled. They use the property that redshift 4 quasars appear much brighter in images taken though a red filter than through a blue filter; the difference is large enough to distinguish the quasars from the very large number of cool stars that are present on images of the sky at faint limits. They have covered a much larger area of the sky, more than 2000 square degrees, at brighter magnitudes in their program than have WHO and SSG. They believe that the bright (and therefore intrinsically luminous) quasars they find do not show a decrease in space density at redshifts of 4 or greater. However, WHO note that they compare to redshift 2. When the WHO increase between redshift 2 and 3.2 is taken into account, WHO believe the results are not inconsistent; that is, there is a peak at redshift 3.2.

It is clearly necessary to establish the nature of the apparent peak in space density of the distant quasars. We await the final results of SSG and calibration by Irwin *et al.* of their data.

An important alternative to consider in interpreting the peak in the space density of quasars is the effect of absorption by dusty material in intergalactic space along the lines of sight to quasars. Could it be blocking our view of the most distant objects? And, therefore, is the peak in space density not real? J. Heisler and J. Ostriker have shown that the way the Universe expands from high redshifts and early epochs to the present can naturally produce a sharp increase in the amount of absorbing material at redshifts greater than 2–3. Their calculations indicate that a reasonable agreement with the data could well arise from dust in galaxies along the line of sight. If this is the case, then quasars of highest redshift should look different from ones at lower redshift in that a reddening of their spectral energy distribution and related effects of absorption should occur. More recently, Y. Pei, M. Fall, and J. Bechtold have made observations of quasars in which the line of sight traverses a thick cloud of neutral hydrogen, such as would occur in the disk of a galaxy too distant to be seen otherwise. They have found evidence for the presence of dust in the hydrogen clouds, but in a lower amount relative to the gas in the cloud than is seen in our own galaxy. Fall and Pei have gone on to argue that the dust in such clouds could account for the decline in apparent space density described earlier. We may expect further results on this topic in the near future, as new, more sensitive instrumentation is brought to bear on the increasing number of the very distant quasars that are being found.

By the same token, as larger telescopes are built and better detectors become available, similar investigations on the space density of normal galaxies at high redshift will become feasible. They will surely bring crucial advances in our knowledge of the distant and early Universe. We may expect the next thirty years of quasar and galaxy research to be as exciting as the last thirty years.

CHAPTER 4

•

GALAXIES AT THE LIMIT:
THE EPOCH OF GALAXY FORMATION

•

HYRON SPINRAD

The technical ability of astronomers to obtain images and spectra of very faint galaxies has improved greatly over the last decade. Since galaxies are vast collections of gas and stars, they must physically evolve with time. We should be able to directly observe the time-evolution of galaxies by studying very distant systems; the look-back interval corresponding to the most-distant galaxies known in 1992 now approaches 15 billion years (80% of the total expansion age of the Universe)!

The line spectra of these faint galaxies are invaluable for redshift determination and physical study. The realization that Ly α (121.6 nm), formed in neutral hydrogen gas, is a strong emission line in most active galaxies and perhaps normal star-forming galaxies also, has helped us measure much larger redshifts in 1987–92 than was previously possible. Recall that this wavelength is in the ultraviolet; it can be observed only by satellites. But when galaxies are very far away, their Doppler effect shifts this spectral line into the region of the spectrum that we can observe with large telescopes on Earth. The largest redshifts for radio galaxies now approach $z=3.8$. Differing selection effects control which galaxies can be seen/isolated that far away. At least some red galaxies must form at red-shift $z_f>5$ (where the subscript f stands for the epoch of star formation). Selection effects also influence our ability to use the galaxies as probes for understanding the geometry of the Universe. As we shall see, the present infrared Hubble diagram does favor an open (expanding forever) cosmological model.

OBSERVING DISTANT GALAXIES

Galaxies are the primary macro-units of our Universe. Studies of nearby systems, from the Milky Way to typical galaxies in the great clusters like Virgo and Coma, have led us to some understanding of the varied processes which can occasionally affect/modify the evolution of these huge stellar systems today, and perhaps change them significantly with cosmic epoch.

This chapter concentrates upon studies of very distant galaxies. They are important because this research represents our only clues, weak as they are, to the birth of individual galaxies, long ago and far away. Since we can obtain information on the Universe only at the speed of light, galaxies or quasars at great distance are seen as light left them long ago. They are now observed at 'look-back times' that can be a large fraction (>80%) of the total expansion age of the Universe (roughly 15–17 billion years in all).

Of course we always need to remind ourselves that extragalactic astronomers usually cannot measure a distance directly. We only determine a galaxy's Doppler shift or redshift; z is the symbol for redshift, and is defined as $z=\Delta\lambda/\lambda_0$ where λ_0 is the rest or 'proper' wavelength of some spectral features. We have found empirically, that for relatively nearby galaxies, redshift is proportional to distance, as is required by the simplest of the expanding models of the Universe. All the galaxies are receding from one another. An excellent general review on the extragalactic distance scale and the universal expansion can be found in Rowan-Robinson's book (1985).

To be definite about the most distant galaxies is a non-trivial task. Distance always implies faintness, so that galaxies observed at the 'frontier' in 1992 represent a sizeable technical challenge to observational astronomers. One of the most difficult tasks is to select out candidate objects in the first place. The extragalactic sky is full of 'bland,' relatively nearby galaxies. Astronomers still prefer to use the magnitude scale, a logarithmic notation where large positive numerical values of the objects' magnitude infer faintness. The limit of the unaided eye is magnitude 6, or $V=6$ in shorthand to represent a visual brightness (V stands for 'visual'). $V=10$ can be seen in a small telescope, $V>21$ requires a large aperture telescope and an electronic detector. At faint levels there are many galaxies projected on each small surface area of the sky. Koo (1989) finds 28 galaxies/square arc min. or some $10^5/\mathrm{mag}/\mathrm{deg}^2$ to $B=24$th mag. A decisive choice of which galaxy to observe physically is required. Much of the following will concern three selection methods that have met with recent success. They involve (a) the use of strong cm/meter range radio-frequency emission to label 'active galaxies,' and (b), the use of galaxy colors, from the near-infrared through the optical domain to weed out atypical objects. This color selection method has only recently been advanced as a way to locate the farthest (fairly-normal) galaxies, and has not yet reached a mature stage. But it does have a strong potential. The third method (c), requires the location of a galaxy in a rich cluster, as shown in Fig. 4.1.

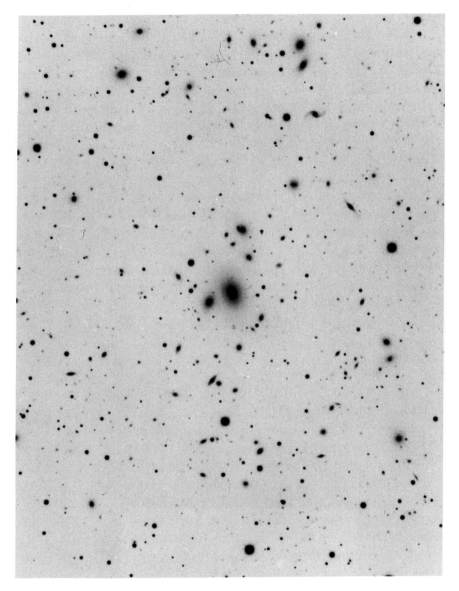

Fig. 4.1. A photograph of the rich and moderately nearby cluster of galaxies called A754. The large galaxy at the center of the cluster would be the object chosen for cosmological study in our context.

Strong radio emission from galaxies is a relatively rare phenomenon, and empirically it has been found to be restricted to massive, luminous elliptical galaxies. A modern general discussion of radio emission from active galaxies can be found in Blanford, Begelman, and Rees (1982). Miley and Chambers (1993) discuss distant radio galaxies from a non-technical point of view. Thus we have signposts to select extragalactic objects of unusually great intrinsic luminosity. These extreme members of the galaxy population can be observed successfully to the largest redshifts. The radio selection method was pioneered by Baade, Minkowski, Sandage, and Gunn and Oke in the U.S., by Ryle and Longair and their students in the U.K., and was adopted by my Berkeley group in 1973. At that time the largest galaxy redshift was that of 3C 295 ($z=0.46$; Minkowski, 1960). We (including Smith, Liebert, Djorgovski, McCarthy, Strauss, and Dickinson) have pushed out the bounds for the galaxies identified with strong radio sources in the 3C catalogue to $z=2.47$ (by 1988). This large redshift has been surpassed by Lilly (1988) for the galaxy associated with the 1 jansky (Jy) radio source 0902+34 ($z=3.39$) and by Chambers *et al.* (1988; 1990) who have observed *at least* one 4C radio galaxy, selected on the basis of an unusually steep radio spectral index, $z=3.8$! (The unit 'jansky' measures radio intensity. A 1 Jy galaxy is a medium-strength radio source, and typically occurs at the distance from us where the number of radio sources exceeds the number expected on a Euclidean distribution by the maximum amount.) The next section will tell the technical story of how this rush towards a distant frontier was accomplished (and still continues).

MODERN METHODS FOR FAINT RADIO GALAXY OBSERVATION

To locate an optically-faint radio galaxy in the presence of more than about 20 other faint galaxies in a one sq min angular field requires an accurate radio position. In practice, a radio synthesis map with a resolution ~1" will usually do the job well, if it has sufficient dynamic range to show weak radio hotspots and extended emission of low surface brightness as well as the classic bi-lobed radio structures. Detection of a central, weak radio core can be crucial to a proper optical detection, especially if the angular size of the classic-double radio source is larger than 30" or so. Fortunately, for many of the most-distant galaxies, the sources are smaller than 30" in their largest angular extent, and we can look directly at the radio center for the active optical object.

The best contemporary radio maps are obtained with the Very Large

Array (VLA), the Westerbork array in the Netherlands, and the MERLIN or the 5-km arrays in the U.K. The Australian radio synthesis telescope is now opening the deep southern hemisphere to detailed long-wavelength scrutiny. Ironically, because of the oversubscription of the fine VLA facility in New Mexico, we have often started our optical work before adequate radio maps of the candidate regions were available. This is certainly not the most-efficient way to proceed, but it can happen.

The next step would be deep CCD imaging in the optical window to locate the candidate object and crudely to define its morphology. It is just becoming possible, in the early 1990s, to obtain images in the near-infrared, at 1.6 and 2.2 μm wavelengths, also.

CCDs, those little solid-state silicon chips, have become the primary detectors of visible light for extragalactic astronomers of this decade. For our project the relatively small size of the CCDs available is not too great a handicap, and due to their linearity, high quantum-efficiency, and large dynamic range, they can be used to image galaxies to roughly $V=24$–25 mag. in about 30 minute integrations with 4-m class telescopes. Are the likely very distant galaxies within this apparent magnitude range? Yes, of course they have been found empirically to be; also theoretical models of galaxy evolution suggest giant galaxies at $z>2$ to still be brighter than our 24th mag. level, *if* any star-formation activity lingers over the first 1–2 Gyr of galaxy life (1 Gyr=1 000 000 000 years).

So, again in a perfect world, deep CCD images in the visual and near-IR domain using photometric bands of U, B, V, R, and perhaps I or K (2.2 μm) would be very desirable. Such a series of images would be a time-consuming task to obtain, but of course would yield much valuable photometric information (cf. Lilly, 1988, 1989). This data is necessary to evaluate the evolutionary state of the stellar content of distant galaxies. Figure 4.2 shows a red bandpass CCD image of the distant radio galaxy, 3C 267, as obtained with the Kitt Peak 4-m reflector and a TI 800×800 CCD array.

A few red galaxies are actually easier to detect in the near-infrared, than in the visible bands. This is due to an intrinsically steep spectrum, the redshift itself, and the improvement of near-infrared detectors (1.6–2.2 μm). Imaging at 2 μm will be an important addition to our battery of techniques, through the 1990s.

Then we are ready for the most demanding empirical task – low-resolution spectroscopy of the galaxy candidate for a redshift determination. Spectroscopy of the faintest galaxies, $V \gtrsim 23$, is a demanding technical chore, because the galaxy is much fainter than the foreground glow of the night sky.

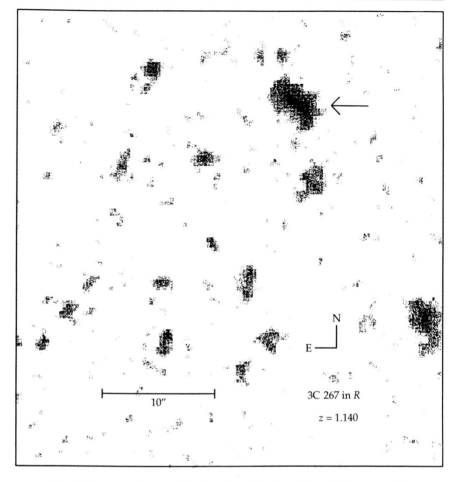

Fig. 4.2. A grey-scale reproduction of a red-region direct CCD image of the radio galaxy associated with 3C 267. The images were obtained at Kitt Peak with the 4-m Mayall reflector. At the redshift $z=1.14$, an arcsec in angular scale corresponds to about 10 kpc linear dimension with the cosmological parameters favored in this paper. Thus, the bi-modal and elongated stellar-light structure shown for 3C 267 has large major axis dimension of 40 kpc.

Thus the signal/noise ratio for the galaxy continuum is going to be quite poor, even with the large telescopes anticipated in the future.

Luckily, most strong radio galaxies have ionized gas in their nuclei and well outside it; this 'thermal' gas displays an emission line spectrum, with the strongest lines usually being Ly α 121.6 nm, C IV 154.9 nm, He II 164.0 nm, C III 190.9 nm, and especially [O II] 372.7 nm. Figures 4.3 and 4.4 illustrate the low-resolution spectra of radio galaxies 3C 55 at $z=0.73$,

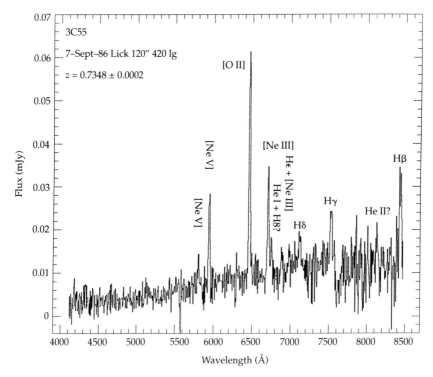

Fig. 4.3. A low-resolution spectrogram of the radio galaxy 3C 55, showing (mainly) the strong and relatively narrow forbidden emission lines of [O II] and [Ne III] and [Ne V] over a weak continuum. This is a high-ionization case, but otherwise typifies the optical (red shifted) spectra of moderate-distance radio galaxies, $0.5 \leq z \leq 1.0$. The radiated power in the the [O II], λ_0 3727 line is large.

showing the near-UV/visible spectrum, and 3C 256 (McCarthy 1988) at z=1.82, showing the emitted deeper UV region, with Ly α the strongest line. (Ly α emission is the result of the H atom changing its energy from the 2nd electronic state down to the 1st and lowest state; since hydrogen is by far the most abundant species this line can be much stronger than other spectral diagnostics in the ultraviolet or elsewhere in the accessible spectrum.)

As we see above the radio galaxies often have a rich emission line spectrum, with high and low ionization species both visible. The great strength of Ly α in most of the high redshift galaxies allows us to obtain reliable redshifts for even the faintest of radio galaxies at $z \gtrsim 1.7$ (which is the rough redshift limit, allowing the line to be visible above the Earth's ozone-horizon, a wavelength of 310 nm or so.) Since the initial detection of Ly α in the spectra

83

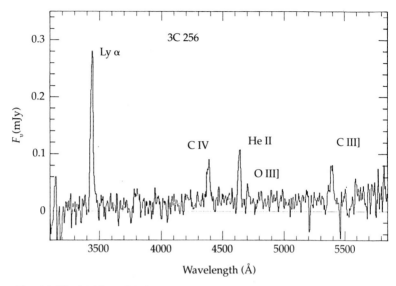

Fig. 4.4. The highly-redshifted UV spectrum of 3C 256 (z=1.82), showing the great strength of Ly α (λ_0 1215.7Å), and the moderate C IV, He II, and C III ions. This spectrogram was also obtained at Lick Observatory by P. J. McCarthy and the writer. This is a relatively bright and 'normal' Ly α galaxy.

of 3C 256 and 239 (Spinrad *et al.* 1985) roughly 30 radio galaxies (mainly 3CR sources, that is, from the 3rd Cambridge Revised Catalogue, but a few significantly weaker ones from the Parkes Selected Areas, the MIT–Greenbank [MG] and the Leiden–Berkeley Deep Survey [LBDS]) have been observed to show moderate to strong Lyα emission. No radio galaxy at z>1.6 for which Ly α could be detected has failed to show this ubiquitous emission line.

MORE ON THE MOST-DISTANT GALAXIES

Indeed it has been the 'Ly α galaxies' which have carried us out to nearly the 'realm of the QSOs' in redshift in the last 2–3 years. There are apparently four types of Ly α galaxies which we can study. It is too early to say whether any of these catagories overlap or are redundant. The culmination of our observations of the ~200 3CR galaxies and the beginning of contemporary studies of the numerous, less-powerful radio emitters culled from later radio surveys have now extended the research potential of this field. But full payoffs may still be years away.

The empirical categories of the Ly α galaxies (Spinrad 1989) are as follows: (a) 'Normal,' large and distant radio galaxies with developed continuum structures (e.g., the galaxies appear concentrated in profile) although their shapes are often non-round or multimodal, (b) 2–3 'blobbier,' less-symmetric objects with weaker or unconcentrated starlight continua, but still showing strong Ly α emission over large area (projected major axis ≈100 kpc). These may well be galaxies in dynamical formation. The third class is (c), the rare radio-quiet companions to distant quasi-stellar objects (QSOs), and the newest category, (d) very recently detected weaker Ly α emitters associated with an intervening galaxy causing a damped Ly α absorption system, or a companion to the 'damping disk'. These may or may not be normal spirals, seen in their youth; the presence of CIV and HeII emission lines in the best-studied companion (Lowenthal *et al.*, 1991) could indicate a weak, active nucleus in that radio-quiet galaxy. The first three classes were described in detail in detail by Spinrad (1988, 1989).

The images of the relatively common type (a) normal Ly α galaxies are quite diverse in shape and in isophotal size; Fig. 4.5 shows 3C 368, a proto-typically aligned galaxy with z=1.14. This alignment is generally present for both continua (starlight) and the Ly α line images (McCarthy *et al.* 1989, McCarthy 1988). The emitted ultraviolet light comes mainly from the hottest stellar population. This probably represents a minority contribution to the

3C368 (B + V + R continuum)   Levels: 8%   15%   25%   50%   80%

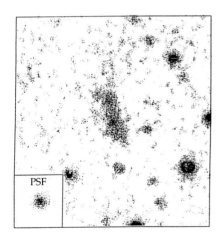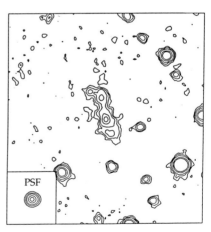

Fig. 4.5. The composite optical image of 3C 368, the prototype for elongated and aligned radio galaxies. Both grey-scales and isophote contours are pre-sented here. The main axis of the stellar image is, as shown, N–S (pa≈10°) as is the radio source structure. There is evidence for substructure in the contours (right) of this image. From Djorgovski *et al.* (1987). The angular scale over the galaxy is about 8″ or ~ 80 kpc (z=1.14). PSF is the point-spread function, the image of a point.

galaxy mass. However, infrared imaging of the visually-elongated radio galaxies at z≥1 at 2.2 μm (Chambers, Miley, and Joyce 1988; Eisenhardt *et al.* 1989) also shows a similar elongation. Thus the 2 μm (1 μm emitted) light coming from the arguably oldest (presumably an original) stellar population shows that it is also not dynamically relaxed. These distant radio galaxies have a basic morphology greatly different from nearby large E galaxies, radio sources or not. The distant radio-emitting systems show a strong, presumably causal, relationship of their optical shapes to their radio source morphologies. They are generally aligned along the radio axes (McCarthy *et al.* 1987 and 1989). Is this due to most of the continuing star-formation being triggered by the interaction of the outflowing radio jet and its early gaseous environment? Do weaker radio sources or radio-quiet young galaxies also have preferred formation axis? That is a key future question, I believe. Recent optical and near-infrared imaging by Rigler *et al.* (1992) has provided evidence that the star formation probably triggered by the radio source ejection may be a moderate perturbation to an older, more-relaxed underlying stellar population that has been 'in place' for some time. The radio source may pro-

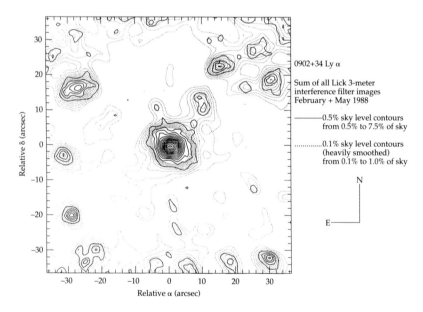

Fig. 4.6. Deep isophotal contours in the Ly α line for the distant radio galaxy 0902+34 (z=3.4). Again, 1″ corresponds to about 10 kpc, so the outer isophotes of the giant Ly α cloud extend across a *diameter* of some 300 kpc, or more. Observations by Dickinson, Spinrad, and McCarthy at Lick Observatory in 1988.

duce additional, aligned star-formation in the interstellar (or circumgalactic) gas available to moderate-redshift large radio galaxies, but would not be morphologically indicative of the overall mass-distribution of the galaxy. Continued imaging in the optical and infrared with high angular resolution with the repaired Hubble Space Telescope is an obvious desideratum.

Some of the observable Ly α structures of our 'class (a)' systems (in the ionized gas itself) are especially interesting. Four hours of narrow-band (special filter) imaging in the Ly α line on the 1 Jy galaxy 0902+34 (z=3.4) has been used to form the composite Ly α image shown as Fig. 4.6. The hydrogen envelope of 0902+34 is truly enormous; it can be traced to a full 15″ in radius, roughly 150 kpc! But in fact, it must be even larger, as we recall that line surface brightness is cosmologically dimmed by a $(1+z)^4$ term! To my surprise, the azimutually-averaged surface brightness profiles of both distant 0902+34 and the forming galaxy 3C 326.1 (McCarthy *et al.* 1987) are exponential in their radial brightness decrease. This form resembles the surface brightness distribution seen in nearby *disk* galaxies, not the supposed giant eliptical systems we've imagined were to be the final product of radio galaxy evolution. The possible physical connection is not at all clear. The spectrum of the 1 Jy galaxy 0902+34, with the dominant Ly α line strikingly strong, is shown as Fig. 4.7.

The second type of Ly α galaxies are (b) the immature and probably forming systems 3C 326.1 and 3C 294. Both these strong radio sources lie at z=1.8. The main difference between these two radio 'galaxies' and the majority of Ly α galaxies of type (a) is that 3C 326.1 and 3C 294 do not have concentrated continua; there is no obvious mature-looking 'parent' galaxy or nucleus for the radio source or the gas-ionization visible in optical (emitted ultraviolet) wavelengths. However, Lilly has found an infrared peak that could be the main radio source and 'dominant' center in his 2 μm images of 3C 326.1, while McCarthy *et al.* (1990) suggest 3C 294 *may* also show an infrared peak near its radio core. Still the ultraviolet light is faint and 'splotchy' – very reminiscent of an *un-organized* collection of sub-units, rather than a mature, large galaxy of stars. Baron and White (1987) favor a theoretical model of primeval galaxies where the bulk of star formation occurs 'late', $z \geq 2$ or so. Their model is also clumpy and has a substantial (tens of kpc) extent during the dissipative collapse era – and superficially does resemble 3C 326.1.

Observationally 3C 326.1 (McCarthy *et al.* 1987) and 3C 295 are mainly visible as giant Ly α clouds, 100 kpc or so in length.

These hydrogen gas clouds overfill the (normal-appearance) radio source morphologies. But there is still an orientation effect, e.g. the gas and weak

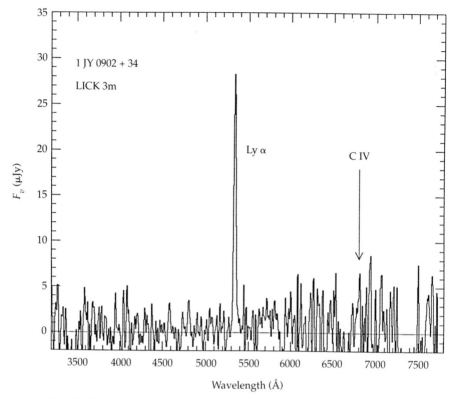

Fig. 4.7. The averaged spectrum of the galaxy 0902+34 (central regions), showing the very strong Ly α and detectable C IV emission lines. There appears to be a (extrinsic) Ly α forest below the emission line, but better spectra with better signal-to-noise ratios are required to firm up that conclusion. Reductions by McCarthy and Spinrad.

continuum blobs of our putative protogalaxies are still aligned along the radio source axis.

3C 294 differs quantitatively from 3C 326.1 in that the organized velocity field of the former protogalaxy is exceptionally large. In 3C 294 the total Ly α velocity range is 1200 km s^{-1}, while the Ly α profile at the cloud center has σ=700 km s^{-1}, a large velocity dispersion also. The 3C 326.1 emission region is considerably more quiescent.

One other difference stands out: in the radio core region of 3C 294 the optical/UV line spectrum shows an area of unusually high gas ionization, with spectral lines of N V, C III, C IV, and He II (in addition to Ly α). These high-ionization species are not seen elsewhere in 3C 294 or at all in 3C 326.1. The source of photoionization at the core of 3C 294 is uncertain. Hot stars (the

source of photoionization at the core of 3C 294 is uncertain. Hot stars (the just-detectable ultraviolet knots) could produce enough ionizing photons to yield the observed Ly α line luminosity [$J(\alpha)=5\times10^{44}$ ergs s^{-1}], but a source of 'harder' ultraviolet radiation than is provided by O stars may be needed for the highly ionized species, like C IV. Probably a buried or anisotropically radiating Active Galactic Nucleus (AGN) is required.

If OB stars ionize the 3C 294 hydrogen, the Ly α line and Lyman continuum energy requires a total star formation rate exceeding 300 $M_\odot$ yr^{-1} (300 times the mass of the Sun each year) – a rate that will exhaust the interstellar matter of even a large forming galaxy quite rapidly. The gas cannot even last a billion years if stars (normal luminosity function assumed) are formed at 300 $M_\odot$ yr^{-1}. So 3C 294 may be a very young system; still it has managed to grow a nuclear black hole (presumably responsible for the central radio source) and produce some nitrogen and carbon, presumably through stellar nucleosynthesis. The total situation does not fit a simple pattern.

The third type of Ly α galaxy (c) is quite rare; only two or three bonafide Ly α companions to more luminous QSOs have been discovered. Osmer's chapter in this book discusses distant QSOs; a general article on 3C 273, the nearest luminous quasar, can be found in Courvoisier and Robson (1991). Galaxies at great distance from us may occasionally be physically related to quasars of known redshift (their active 'big-brothers'). At z~3 surveys for Ly α emitting *companion galaxies* have been carried out by Djorgovski *et al.* (1985, 1987), Hu and Cowie (1987), and Elston (1989), while a serendipitous detection has recently been made by Steidel, Sargent, and Dickinson (1991). The original QSO companion to PKS 1614+051 (Djorgovski *et al.* 1985) may be a somewhat 'active-nucleus' object, while the Steidel companion has only weak and narrow Ly α emission; it's not an active galactic nucleus. Thus QSO companions, physically associated but also separate galaxies from their 'big brothers' are rare – perhaps more rare than one would naively predict. What about companions to radio galaxies (presumed giant elliptical-galaxies precursors)? So far, at *large* (z>2) redshifts no Ly α companions are known to this writer. Dickinson and Spinrad searched unsuccessfully for putative Ly α emitting companions to the distant 1 Jy radio galaxy 0902+34 (Lilly; z=3.39). None with emission line flux greater than 2 per cent of the main radio galaxy were detected. The limit on the companions, $J(\alpha)\leq1.6\times10^{43}$ ergs s^{-1}, is an emission measure which (in the absence of dust) yields a limit to the star-formation rate of ~10 $M_\odot$ yr^{-1}. One wonders if any early-life companions have already (at z>3) been integrated into the main, large galaxy (after only 1–2 formation rate comparable to our Milky Way's rate (~5 $M_\odot$ yr^{-1}) would be very difficult to detect at z=3 by our techniques.

While about 20 of the so-called damped absorption systems at z>2 have been recognized, no Ly α emission from the presumed intervening galaxy had been detected until 1988. Finally, Hunstead and Pettini (1989) succeeded in observing the weak emission component in a z=2.5 intervening damped Ly α system. Their detection was difficult and so far yields little information on the spatial extent of the intervening system. It has a narrow velocity width and probably a fairly small size (9×18 kpc at maximum). Perhaps the intervening gas-rich matter is a portion of a dwarf galaxy. The conclusion is also interesting quantitatively, as the inferred star formation rate from the Ly α line intensity is about 3 $M_\odot$ yr^{-1} (for q_0=0), and that implies a time scale of about a Gyr (10^9 yr) to turn most of the assumed gas disk into stars. Obviously, more data on this damped system and others will be needed to firm up the above hypothesis; the best point about the 'category 4' Ly α emitter is that its selection was unbiased toward galaxy 'activity,' depending only upon random lines of sight through occasional gas-rich environs.

Finally, we discuss a conceptual method and first results which suggest using the likely robustness of the Lyman limit edge at rest 91.2 nm as a large discontinuity in galaxy spectra. Cowie and Lilly (CL) (1989), Spinrad, and most-recently Steidel and Hamilton (1992) have suggested utilizing photometric criteria to advance intrinsic Lyman-limit candidates at z>3 (see Table 4.1 for some details on the optical placement of the Lyman discontinuity at several redshifts). The physical point is that even very hot OB stars probably have 91.2 nm discontinuities of about a factor 3; any neutral gas covering the starlight (from our direction) will add optical depth to the radiation harder than the limit – perhaps making the galaxy 'completely-dark' at λ_0<91.2. *Extrinsic* limit systems, with $z_a \leq z_{gal}$ could (statistically) increase the amplitude of the broad-band limit discontinuity. Galaxies with some active star formation should be blue at rest wavelengths >150.0 nm; at z~3 the ultraviolet (*U* band) will be depressed relative to the *B* and *V* bands; thus a *U*-faint, (*B-V*) blue galaxy will be a qualitative Ly-limit, high-z candidate. Perhaps our own Milky Way Galaxy resembled the young, star-forming system we imagine here, in its youth. An interested reader may wish to refer to a modern review on the evolution of the Milky Way by Van den Bergh and Hesser (1993). Steidel and Hamilton have quantified this photometric approach for a specific redshift (that of the QSO-Lyman limit system, Q0000-263 at z=3.4). They identify one or perhaps two *very faint* galaxies found on the basis of *U*-band depression (upper limits, only) as candidates for a common redshift with the limit system itself, and thus if some spectroscopic information can be obtained, we'll have a good future method of locating star-forming galaxies over $3.0 < z \leq 6$. There are certainly plenty of faint blue galaxies in the extant counts to B magnitude 25; one thinks that most of them are nearby, moderate-to-low luminosity galaxies, but some of the 'blue-fuzzies' at very faint levels, $24 < B \leq 26$, ought to be galaxies in their extremely early evolutionary phases (cf., CL), with the initial star formation going on ($z > 2$, perhaps). Of course, as usual in extragalactic astronomy, the difficult 'proof-of-the-pudding' must be a spectroscopic redshift measurement – and it might as well be on a Lyman limit candidate found by multi-color photometry.

Table 4.1 shows observed wavelengths for the onset of the Lyman continuum; 91.2 nm propagates through the U band ($\lambda_{eff} = 360.0$ nm) at $z = 3.0$, so photometry above and below the discontinuity when it's this convenient should eventually become precise enough to locate distant spirals and help model the integrated cosmological ionizing flux from young galaxies. That ultraviolet light could be significant in the maintenance of the high ionization state of the intergalactic medium (e.g. Shapiro and Giroux, 1989). As Table 4.1 shows, mere detection of galaxies at $z \geq 9$ is a task for the infrared.

Table 4.1. *The Lyman limit redshifted to the optical*

Assumed z	(91.2) (1+z)	Photometric band involved	Notes
1.0	182.4 nm	–	Hubble Space Telescope observations required
2.0	273.6	–	Hubble Space Telescope observations required
3.0	364.8	U	As per CL. (1/2 U 'dark')
4.0	456.0	B	U and $^1/_2$ B 'dark'
5.0	547.2	V	U, B, and $^1/_2$ V 'dark'
7.0	729.6	I	U, B, V, R 'dark'; detect at I or K.
9.0	912.0	–	Optical (CCD) window – 'dark'

PHOTOMETRIC EVOLUTIONARY MODELS OF GALAXIES

Another major topic in which faint galaxies may help is the look-back time study of the evolution of a galaxy's population of stars. Photometric models of the evolution of the stellar content of distant galaxies were pioneered by Tinsley (1980) and followed by the research of Bruzual (1983), Renzini and

Buzzoni (1986) and Rocca-Volmerange and Guiderdoni (1988). Observational aspects of the problem have been discussed by Spinrad (1986), Hamilton (1985), and Lilly (1988, 1989). In this chapter I will deal specifically with observations and models that hope to constrain the epoch of initial heavy star-formation in galaxies.

Lilly's observations of the 1 Jy radio galaxies at $z>1$ indicates an important lower bound to the epoch of initial star formation in some large elliptical galaxies. The sharpest constraint, of course, will come from galaxies whose stars already appear old at large redshift ($z>1$). The red (emitted) radiation arises predominantly from the oldest stars present; the question we attempt to answer by stellar population synthesis is quantitative – how old must these old stars be? A minimum age for the dominant initial population is the best we can do in practice, utilizing updates of the popular Bruzual (1983) models. The distant red-color radio elliptical galaxies observed by Windhorst et al. (1986) and by Lilly (1989) have spectral energy distributions still rising toward (emitted) red wavelengths, even though the galaxies are seen as they radiated over 8 billion years ago! Lilly's (1988) original data on the particularly distant 1 Jy galaxy 0902+34 (at $z=3.4$) are now in doubt (Eisenhardt and Dickinson 1992) as far as it showing a dominant red population, plus considerable extra UV light from a young minority population which controls the short wavelengths below emitted λ_0 300 nm. But certainly some distant galaxies have red colors.

The vital near-infrared data for the reddest radio galaxies show the old, 'dead' stars still produce about 90% of the light at a wavelength of 500.0 nm, so the early single-burst type of population model can be safely applied. The synthesis models which best fit the red components of radio galaxies yield an age 1.5–2 Gyrs for the older stars. In the most extreme cases, the 'formation redshift' (really the redshift corresponding to the last epoch of major star formation) is $z_f>10$ for $q_0=+0.5$! So some massive galaxies form very early.

On the other side of the 'observational fence' we note that 3C 326.1 and 3C 294, discussed earlier, may just be forming most of their stars at $z=1.8$. Of course, there is no compelling reason to anticipate coevality! Galaxies may well form over some range in cosmic epoch.

One more uncertain note ends this section: while we think mostly old stars dominate the observed near-infrared light of distant elliptical galaxies, the spatial alignments recently seen along the galaxy radio axes (Chambers et al. 1987; Eisenhardt et al. 1989) at 2 μm for distant radio galaxies *may* suggest a *younger age* population to play some role, even at long wavelengths. That is because the galaxies must dynamically relax to a more spherical shape when they become 'middle-aged'. A quantitative understanding of this alignment phenomenon seems requisite in the near future; Rigler *et al.* (1992) provide a start in this direction.

EXTRINSIC INTERVENING MATTER – MODULATION OF THE DISTANT GALAXY SPECTRUM

Observations of galaxies, not only quasars, are now taking us out to objects sufficiently distant that intervening cosmological clouds may influence the spectral and even the apparent spatial structures of galaxies at $z>3$ or so. The Universe is not quite transparent.

One probably important extrinsic phenomenon is the integrated effect of the Ly α 'forest' (many weak Lyα absorption lines at wavelengths just short of Ly α emission in the galaxy itself). This is seen by careful inspection of the noisy continuum spectrum of 0902+34 (z=3.4), note again Fig. 4.7, and should be present in the 4C galaxies with $z>3$ also (Chambers *et al.* 1988). The local galaxy continuum below λ_0 121.6 is depressed compared to that above 121.6 nm; this Ly forest 'ledge' is well known to observers of QSOs (Steidel and Sargent 1987) for $z \geq 3$. The density of the intervening H I clouds increases rapidly with redshift, so we should anticipate larger Ly α forest discontinuities in the continua of the most distant galaxies. It is, I believe, plausible to hope that the 'forest edge' will be a discriminant for the future *location* of galaxies at $z>4$! In such a scenario, no strong Ly α emission line would be needed for the redshift determination.

Obscuration by *intervening dust clouds* (in galaxy disks, presumably) could also be important in the surface brightness distribution studies of large, distant galaxies which are selected on the basis of radio emission. Ostriker and Heisler (1984), Heisler and Ostriker (1988), and Weedman (1987) have suggested that many lines of sight to distant QSOs may be dust-obscured. Since our sample of distant galaxies is primarily radio-selected, the detected optical objects should fairly sample the transparency of the Universe; we can qualitatively test their suggestions of heavy cumulative absorption for $z>3$ when several large galaxies at these large redshifts are studied in detail. The initial surface-brightness profiles in the Ly α line in 0902+34 (Dickinson and Spinrad, in preparation) show no evidence for dust-path 'holes' in the smooth Ly α profile (see again Fig. 4.6) on the sky. This analysis is, of course, limited by our lack of complete understanding of the underlying light distribution and H-ionization sources in a relatively young galaxy.

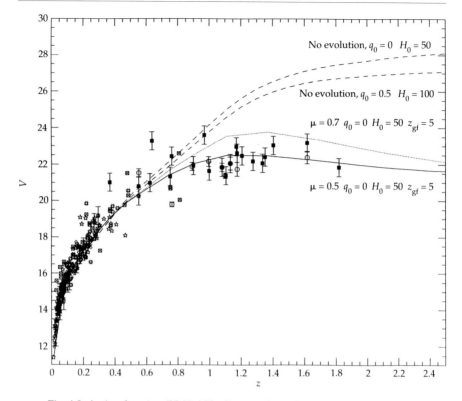

Fig. 4.8. A visual-region (*V*) Hubble diagram for radio galaxies from Spinrad and Djorgovski (1987). Note the large scatter of the plotted photometric points (3C 256 at z=1.82, at that time was the high-z data boundary). Note also the large sensitivity to galaxy evolution (model curves with evolution-lower curves (μ-0.7, 0.5) and without ('no evolution')).

COSMOLOGY WITH DISTANT RADIO GALAXIES IN THE NEAR INFRARED

A principal goal of modern cosmology is to determine the global geometric structure and evolutionary history of the Universe. The Hubble diagram, a plot of the (m, z) relation for a given type of luminous galaxy (here the radio galaxies of high power), has been one of the classic methods in this quest (c.f. Sandage 1975, 1988). But in the real observational world these goals, in particular the determination of the deceleration parameter q_0 and the quantitative evaluation of galaxy luminosity evolution, have been badly intertwined. Thus neither has been determined to satisfactory precision. Our ability to make the raw brightness measurements of faint galaxies has been greatly improved by the CCD and IR (2 μm)-array technologies of this decade [see Spinrad and Djorgovski (1987) and Eisenhardt and Lebofsky (1987) and Lilly (1989).

There still is residual worry about the quality of our radio galaxy 'standard candle'; do they radiate equally at visible and near-infrared wavelengths? Empirically they show a small scatter in infrared (2 μm) magnitude at a given z (Lilly and Longair 1984, Lilly, 1989)]. But of course that's only a portion of the real problem.

The results for the visible region Hubble diagram (see Fig. 4.8) were summarized by Spinrad and Djorgovski (1987); there is much scatter in V mag. at a given redshift beyond $z=0.7$. Presumably this scatter is mainly due to the sensitivity of the emitted ultraviolet portion of the composite stellar spectrum to galactic evolution parameters, like the time-decay of the star-format rate. The non-negligible cosmological dependences in the V Hubble diagram for our most-distant galaxies are masked by the likely stochastic nature of the star-formation history in our radio galaxy sample.

But in the near-infrared the K-band Hubble diagram fares better (see Fig. 9 in Spinrad and Djorgovski and Fig. 4.9, here). At these longer wavelengths, there is naturally less sensitivity to the time-changing evolution of fairly massive stars, so that stellar model evolutionary differences play less of a discriminatory role than they do in the visible (emitted ultraviolet for $z \gtrsim 1$). Only fairly extreme stellar model differences will compromise the utility of the IR Hubble diagram for cosmology. In the infrared Hubble diagram the scatter of the galaxy points is modest, and the cosmological differences between major geometric world models (open v. just closed, for example) are considerable, compared to the 'noise,' for $z \geq 1.0$.

From this picture (Fig. 4.9) we would estimate q_0 to lie between 0.2 and 0.3, if no large and *systematic* problems sway our outlook. This conclusion, if supported, would indicate a still-open Universe with an average density about 40% that demanded for closure. So, if verified, this relation would signal a Universe expanding forever! Future large-scale programs of imaging photometry at the infrared wavelength of the K filter will further extend and improve the infrared Hubble diagram. We are now starting to observe radio galaxies to $K=19$, and that *could* suggest $z\simeq4$, or more. We look forward to a full inclusion of steep-spectrum 4C galaxies (Chambers, Miley and van Bruegel 1988) and 1 Jy galaxies (Lilly 1989) with the MG and 3CR data mentioned here. Then we will be utilizing the potential data on the youngest and

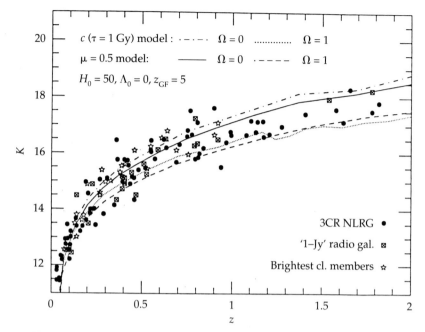

Fig. 4.9. The *K*-band (2.2 μ) Hubble diagram shows much less sensitivity to stellar evolution in the galaxies and should be able to constrain cosmology well. Curves through the observed galaxy data points (from top to bottom) represent open and closed Universe models with slightly different star formation histories. This figure from Spinrad and Djorgovski suggests an open cosmology if the data points and the interpretational models are taken seriously. The graph is plotted for a Hubble contant of 50, a cosmological constant of 0, and the redshift of galaxy formation (z_{gf}) of 50. NLGR stands for narrow-line radio galaxies, a class of radio-loud galaxies with narrow emission lines. The ones plotted here are from the 3rd Cambridge (3C) calalogue of radio sources, as revised (thus 3CR).

most distant of assembled giant galaxies, residents of a younger, active Universe when their light departed on the long journey here. These galaxies should help in attacking one of cosmology's oldest and most-important problems.

•

REFERENCES AND RELATED READING

•

Related reading for Chapter 1:
Observing the farthest things in the Universe

Ferris, Timothy, *Coming of Age in the Milky Way*. New York, Morrow, 1988.

Lemonick, Michael D., *The Light at the Edge of the Universe*. New York, Villard Books, 1993.

Overbye, Dennis, *Lonely Hearts of the Cosmos: The Scientific Quest for the Secret of the Universe*, New York, Harper Collins, 1991.

Pasachoff, Jay M., *Astronomy: From the Earth to the Universe*, 4th edition, 1995 version, Philadelphia, Saunders College Publishing, 1995.

Pasachoff, Jay M., and Donald H. Menzel, *A Field Guide to the Stars and Planets*, 3rd edition, Boston, Houghton Mifflin, 1992.

Silk, Joseph I., *The Big Bang*, New York, Freeman, 1989.

Related reading for Chapter 2:
The cosmic background radiation

Boggess, N. W. *et al.* The COBE Mission: its design and performance two years after launch. *Astrophysical Journal*, **397**, 420, 1989.

Cornell, J. (ed.) *Bubbles, Voids and Bumps in Time: the New Cosmology*. Cambridge University Press, 1989.

Gulkis, S., Lubin, P. M., Meyer, S. S. and Silverberg, R. F. The Cosmic Background Explorer. *Scientific American*, **262**, No. 1, pp. 132 ff. January, 1990.

Harrison, E. R. *Cosmology, the Science of the Universe*. Cambridge University Press, 1981.

Longair, M. S. *The Origins of Our Universe*. Cambridge University Press, 1990.

Pasachoff, J. M. *Astronomy: From the Earth to the Universe*, 4th edition, 1995 version. Saunders, 1995.

Peebles, P. J. E. *Principles of Physical Cosmology*. Princeton University Press, 1973.

Shu, F. *The Physical Universe, an Introduction to Astronomy*. University Science Books, 1982.

Silk, J. *The Big Bang*. W. H. Freeman, 1989.

Weinberg, S. *The First Three Minutes*. Basic Books, 1977.

Related reading for Chapter 3: Quasars

Crampton, D. (ed). The space distribution of quasars. *Astronomical Society of the Pacific.* Conference Series vol. 21, 1991.

Field, G. B., Arp, H., and Bahcall, J. N. *The Redshift Controversy.* Reading, Mass. Benjamin, 1973.

Osmer, P., Porter, A., Green, R., and Foltz, O. (ed.s). *Astronomical Society of the Pacific.* Conference Series vol. 2, 1988.

Swarup, G., and Kapahi, V. (eds) *Quasars. International Astronomical Union Symposium No. 119.* Dordrecht, Netherlands, Reidel, 1986.

References for Chapter 4:
Galaxies at the limit: the epoch of galaxy formation

Baron, E., and S. D. M. White 1987, *Astrophys. J.*, **322**, 585.

Blandford, R. D., M. Begelman, and M. J. Rees, 1982, *Sci. Am.*, **246**, 124.

Bruzual, G. A. 1983, *Astrophys. J.*, **273**, 105.

Chambers, K. C., G. K. Miley, and W. van Breugel, 1987, *Nature*, **329**, 609.

Chambers, K. C., G. K. Miley, and R. R. Joyce, 1988a, *Astrophys. J.*, **329**, L75.

Chambers, K. C., G. K. Miley, and W. van Bruegel, 1988b, *Astrophys. J.*, **327**, L47.

Chambers, K. C., G. K. Miley, and W. van Bruegel, 1988c, talk presented at the 20th General Assembly of the *Intl. Astron Union* in Baltimore.

Chambers, K. C., G. K. Miley, and W. van Bruegel, 1990, *Astrophys. J.*, **363**, 21.

Courvoisier, T. and E. I. Robson, 1991, *Sci. Am.*, **264**, 50.

Cowie, L. L., and S. J. Lilly, [CL] 1989, *Astrophys. J.*, **336**, L41.

Djorgovski, S., H. Spinrad, P. McCarthy, and M. A. Strauss, 1985, *Astrophys. J.*, **299**, L1.

Djorgovski, S., M. A. Strauss, H. Spinrad, R. Perley, and P. J. McCarthy, 1987, *Astron. J.*, **93**, 1318.

Eisenhardt, P., A. Chokshi, and H. Spinrad, 1989, paper presented at the 173rd *Am. Astron. Soc.* Meeting in Boston.

Eisenhardt, P., and M. Lebofsky, 1987, *Astrophys. J.*, **316**, 70.

Eisenhardt, P., and M. Dickinson, 1992, *Astrophys. J. L.*, **339**, L47.

Elston, R. 1989, paper presented at the 173rd *Am. Astron. Soc.* in Boston.

Hamilton, D. 1985, *Astrophys. J.* , **297**, 371.

Heisler, J., and J. Ostriker, 1988, *Astrophys. J.*, **332**, 543.

Hu, E., and L. Cowie, 1987, *Astrophys. J.*, **317**, L7.

Hunstead, R. and M. Pettini, 1989, in *Durham NATO Workshop, 'Epoch of Galaxy Formation,'* ed. C. S. Frenk (Kluwer Academic Press), p. 115.

Koo, D., 1989, in *Durham NATO Workshop, 'Epoch of Galaxy Formation,'* ed. C. S. Frenk (Kluwer Academic Press), p. 71.

Lilly, S. J. 1988, *Astrophys. J.*, **333**, 161.

Lilly, S. J., 1989, in *Durham NATO Workshop, 'Epoch of Galaxy Formation,'* ed. C. S. Frenk (Kluwer Academic Press), p. 63.

Lilly, S. J., and M. S. Longair, 1984, *Monthly Notes Royal Astron. Soc.*, **211**, 833.

Lowenthal, J. D., C. R. Hogan, R. F. Green, A. Caulet, B. E. Woodgate, L. Brown, and C. B. Foltz, 1991, *Astrophys. J.*, **377**, L73.

McCarthy, P. J. 1988, University of California Ph.D. Thesis.

McCarthy, P. J., H. Spinrad, S. Djorgovski, M. A. Strauss, W. van Breugel, and J. Liebert, 1987, *Astrophys. J.*, **319**, L9.

McCarthy, P. J., H. Spinrad, and W. van Breugel, 1989, *Intl. Astron. Union Symposium*, **134**, Santa Cruz, eds. D. E. Osterbrock and J. Miller.

McCarthy, P. J., H. Spinrad, W. van Bruegel, J. Liebert, M. Dickinson, S. Djorgovski, and P. Eisenhardt, 1990, *Astrophys. J.*, **365**, 487.

Miley, G. K., and K. C. Chambers, 1993. *Sci. Am.* **268**, 6.

Minkowski, R. 1960, *Astrophys. J.*, **132**, 908.

Ostriker, J., and J. Heisler, 1984, *Astrophys. J.*, **278**, 1.

Renzini, A., and A. Buzzoni, 1986, in *Spectral Evolution of Galaxies*, eds. C. Chiosi and A. Renzini (Dordrecht: D. Reidel), p. 95.

Rigler, M., S. J. Lilly, A. Stockton, F. Hammer, and O. LeFevre, 1992, *Astrophys. J.*, **385**, 61.

Rocca-Volmerange, B., and B. Guiderdoni, 1988, *Astr. and Astrophys. Suppl. Ser.*, **75**, 93.

Rowan-Robinson, M. 1985, *The Cosmic Distance Ladder*, W. H. Freeman & Co., New York.

Sandage, A. R. 1975, in *Galaxies and the Universe* vol. IX of *Stars and Stellar Systems*, eds. A. Sandage, M. Sandage, and J. Kristian, p. 761.

Sandage, A. R. 1988, *Ann. Reviews Astron. and Astrophys. J.*, **26**, 561.

Shapiro, P. R., and M. L. Giroux, 1989, in *Durham NATO Workshop, 'Epoch of Galaxy Formation'*, ed. C. S. Frenk (Kluwer Academic Press), p. 153.

Spinrad, H. 1986, *Publ. Astron. Soc. of the Pacific*, **98**, 269.

Spinrad, H. 1988, in *Institut d'Astrophysique Conference 'High Redshift and Primeval Galaxies,'* eds. J. Bergeron, D. Kunth, B. Rocca-Volmerange, and J. Tran Thanh Van (Editions Frontières, Paris), p. 59.

Spinrad, H. 1989, in *Durham NATO Workshop, 'Epoch of Galaxy Formation,'* ed. C. S. Frenk (Kluwer Academic Press), p. 39.

Spinrad, H., A. V. Filippenko, S. Wyckoff, J. T. Stocke, R. M. Wagner, and D. G. Lawrie, 1985, *Astrophys. J.*, **299**, L7.

Spinrad, H., and S. Djorgovski, 1987, in *Intl. Astron. Union Symp.*, *124 'Observational Cosmology,'* eds. A. Hewitt, G. Burbidge, and L. Z. Fang, p. 29.

Steidel, C., and D. Hamilton, 1992, *Astron. J.*, **104**, 941.

Steidel, C., and W. L. W. Sargent, 1987, *Astrophys. J.*, **313**, 171.

Steidel, C., W. L. W. Sargent, and M. Dickinson, 1991, *Astrophys. J.*, **101**, 1187.

Tinsley, B. M. 1980, *Astrophys. J.*, **241**, 41.

Van den Bergh, S. and J. E. Hesser, 1993, *Sci. Am.*, **268**, 72.

Weedman, D. 1987, in *NASA Conf. Public. 2466 'Star Formation in Galaxies,'* p. 351.

Windhorst, R., D. C. Koo, and H. Spinrad, 1986 in *Galaxy Distances and Deviations from Universal Expansion*, eds. B. F. Madore and R. B. Tully, p. 97.

INDEX

References to illustrations, either photographs or drawings, are in italics. References to tables are followed by t.
Significant initial numbers followed by letters are alphabetized under their spellings.
Less important initial numbers are ignored in alphabetizing, e.g.,
3C 273 appears at the beginning of the Cs and M13 appears at the beginning of the Ms.